COLLINS WILD GUIDE

NIGHT SKY

URSA MINOR

Storm Dunlop

Charts and diagrams by
Wil Tirion

HarperCollins*Publishers*

HarperCollins*Publishers*
77–85 Fulham Palace Road
London
W6 8JB

00 02 04 03 01 99

2 4 6 8 10 9 7 5 3 1

ISBN 0 00 220127 5

All photographs by Steve Edberg apart from those that
appear on the following pages:
Denis Buczynski: 130; Storm Dunlop: 16, 20, 30; David
Gavine: 155, 157, 141; Peter Gill: 21, 165; Alan Heath: 111;
Robert McNaught: 158; Martin Mobberley: 163

Designed by Liz Bourne

Colour origination by Colourscan, Singapore
Printed and bound by Rotolito Lombarda SpA,
Milan, Italy

INTRODUCTION

The original inspiration for this book came from the comments made by various groups of students attending adult-education classes in astronomy. Many beginners found that when looking at the night sky from their back garden or from their living-room window, they became confused by 'so many stars'. A few practical sessions out in the college grounds, with some guidance about how to locate and recognize different constellations, and everything began to fall into place. Soon, they were not only picking out the constellations but confidently identifying planets, and engaging in friendly competitions as to who could see most stars in the Pleiades or the greatest detail on the Moon with their binoculars.

Strangely enough, although there are many excellent books on astronomy, often with detailed descriptions of the various constellations, there are very few – and those all out-of-print – that really catered for beginners by showing them how to progress from no basic knowledge of the sky to being able to recognize all the constellations visible in the northern sky. This is what this book sets out to do. You don't need any equipment at all, although a pair of binoculars would be useful. Try it. You will probably surprise yourself by how quickly the patterns of stars become familiar.

Naturally, there are limitations on what can (or should) be included in a book such as this one. Astronomy is a hobby that may be pursued on many different levels, from simply enjoying the spectacle of the night sky, to serious amateur scientific work. This book confines itself to objects that may be observed with little (or no) optical equipment. Let us hope that you will find it a gateway to a lifetime interest in astronomy, which is not only the oldest, but still probably the most fascinating, of all the sciences.

HOW TO USE THIS BOOK

This books aims to help you find your way around the sky, so that you are able to recognize the different constellations. It will also help you to see some of the brightest of the interesting objects in the night sky. The first section (pp.10–17) acts as a general introduction to naked-eye and binocular astronomy and gives some general advice on observing and simple photography. This is followed by a short explanation (pp.18–29) of some basic astronomy that will help you to get started.

If you are just beginning, the first thing to do is to learn the constellations that are visible on any clear night. There are just five important ones, and they are described on pp.30–35. Most constellations are visible for only part of the year, so the next section (pp.38–109) gives a series of charts, two for each month, one showing the sky looking north, and the other looking south. Depending on the time of year when you start observing, you can begin at the appropriate month. Each month there is also a detailed description of how to find two or three different constellations. If you follow the charts throughout a whole year, you will have learned how to find all the constellations that are easily visible from these latitudes.

The Moon is an easy object to observe, so maps of various phases are given on pp.114–127. Because planets move from one constellation to another, they can cause confusion, even to experienced astronomers. So the next section (pp.134–53) contains a series of maps for several future years that enable you to locate individual planets. Just look at the pages for the current year to see where the planets are in the sky.

More general details of the different types of events or phenomena that you may see in the night sky are given next (pp.154–167). The final section (pp.168–251) describes individually each constellation visible from the northern hemisphere, with brief details of some of the most interesting objects that may be seen. On these pages, the light-coloured portion of the calendar bars indicates when the particular constellation is easily visible. A key to the symbols used in the constellation charts themselves appears on p.167.

CONTENTS

THE NIGHT SKY

On a dark night when there is no Moon, and far away from city lights, the sky appears to be full of thousands of stars, scattered seemingly at random across the heavens. Some people are overawed by the sheer number of stars that they can see. Learning to identify any of them seems to be an impossible task. In fact, appearances are deceptive, because only about 2000 stars are visible at any one time, even under the most favourable conditions. You certainly do not need to be able to recognize every one. It is not difficult to pick out distinctive patterns among the brighter stars, and this is all you need to find your way around. After quite a short time you will also be able to recognize many of the fainter stars as well.

Ever since ancient times, people have given names to various groups of stars, and the system used world-wide today is one that that has evolved over the centuries. It contains elements introduced by earlier Babylonian, Greek, Roman, and Arab astronomers. There used to be considerable variation in the areas of sky known by particular names, but now, by international agreement, the whole sky is divided into 88 **constellations** with fixed boundaries. Similarly, the various groups of stars were originally associated with, or named after various mythological beings, animals or, later, particular objects. Again by agreement, the Latin names of the constellations are recognized by astronomers world-wide, so they are used in this book. Some stars within an individual constellation may form particularly distinctive patterns, and these are known as **asterisms**. The familiar group of stars known as the **Plough** is actually an asterism, because it forms part of the much larger constellation of Ursa Major. The seven stars of the Plough are our starting point for finding our way around the sky.

The constellations vary greatly in size. Some sprawl across large areas of sky, while others are much smaller. Some contain several bright stars, while others are faint and difficult to detect. This book concentrates on the most important constellations, which are generally easy to find. In addition, because it is intended specifically for northern hemisphere beginners, it omits constellations that may be seen only from the southern hemisphere.

Scattered around the sky are denser collections of stars known as **clusters**. A few are visible to the naked eye, and many more may be seen with binoculars. From a really dark site it is possible to see the Milky Way, a broad, irregular band of light stretching right round the sky. It actually consists of millions of stars, which appear packed closely together where we are looking through the broad disk of the Galaxy. Just one other galaxy is visible to the unaided eye (or two if you are extremely sharp eyed and have exceptionally dark surroundings), but binoculars reveal others as faint smudges of light.

There are many other objects in the night sky. The **Moon**, going through its phases once a month, is the most

At Full Moon, ray craters are the most conspicuous features

conspicuous, but there are also the five **planets** that are visible to the naked eye, and which have been known from antiquity. Of the remaining four planets, one, Uranus, is just about visible with the unaided eye under good conditions, and Jupiter's four main **satellites** may be seen readily with binoculars.

The Moon's orbit sometimes carries it into the Earth's shadow and gives rise to a **lunar eclipse** (p.130). Contrary to what one might expect, there are great variations between different eclipses, so each one is worth watching and leaves a different impression. Although occurring in the daytime sky, **solar eclipses** (p.128) also show great variations, and the chance to observe one is not to be missed.

Then there are **meteors**, the so-called 'shooting stars' that fleetingly race across the sky, and which may sometimes occur in considerable numbers. Although most are faint, on rare occasions one may be bright enough to light up the whole sky. By contrast, **comets**, to which many meteors are related, normally move slowly against the starry background. Some, like Comet Hyakutake in 1996 and Comet Hale-Bopp in 1997, may become very conspicuous when they grow extensive tails.

Closer to the ground, the **aurora** occurs in the high atmosphere and produces wonderful displays that are often particularly spectacular during the long dark nights of winter. In summer, favourably placed observers may be lucky enough to see **noctilucent clouds**, the highest of Earth's clouds, shining in the north around midnight. Apart from all these natural phenomena, however, there are also man-made objects: the **artificial satellites** that orbit the Earth, and which may be seen on any night.

SIMPLE TIPS FOR OBSERVING

It may seem obvious, but because you observe at night, you need to keep warm, so wear adequate clothes. Similarly, try to avoid standing on grass (wet grass, in particular), because your feet quickly become chilled. It has been said, only partly in joke, that good astronomers wear two pairs of socks. In addition, 25% of body heat is lost through the head, so a hat may be just as important.

You also need a dark observing site. In towns and cities, this may be difficult to arrange, but try to find a position where no lights shine directly onto you. The light pollution that nearly everyone suffers, particularly if they live in an urban area, drowns out the fainter stars. The only slight positive advantage is that beginners may find it easier to find their way around, because only the brightest stars are visible.

If possible, give yourself sufficient time for your eyes to adapt to the dark. This dark-adaptation is not the more-or-less instantaneous change in the size of the pupil of the eye that occurs when you move from light to dark, or vice versa, but a growth in sensitivity that takes place over a period of 15–20 minutes or more. Provided the eye remains in the dark, it gradually becomes better at detecting faint objects. In this state the eye is most sensitive to green light, and least sensitive to red. Astronomers always use a low-power red light to illuminate their charts, so that dark-adaptation will not be lost. A torch covered in red plastic or red cellophane will do perfectly well.

Observing constellations high overhead is always difficult. If it is summertime and you want to lie on the grass, that's fine, but normally it is better to use an adjustable garden chair. One with arms is particularly useful if you want to use binoculars, which are discussed in more detail on pp.12–15.

Estimating distances on the sky

It is often useful to be able to make rough estimates of distances on the sky, especially in the initial stages when you are beginning to find your way around. A few people find it difficult to match charts to the sky, because of the differences in scale, and nearly everyone overestimates the sizes of constellations on the sky.

The simplest method is to use your hand, held at arm's length. (Any slight differences between individuals is not important for our purposes.) A finger is slightly more than 1° across, and that is about twice the size of the full Moon. The width across the knuckles is about 7°, the overall width of a clenched fist about 10°, and the span over outstretched fingers is about 22°. You can also hold a ruler graduated in centimetres at arm's length, when 1 cm is approximately 1°. In the dark, however, it is easier to use a hand.

The stars of the Plough also provide a guide. The distance between the Pointers is about 5½°, that between the top stars of the 'bowl' is 10°, and the Plough's overall length is 25°.

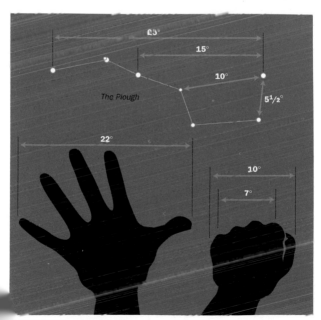

Estimating angles on the sky

CHOOSING AND USING BINOCULARS

Beginners often imagine that astronomy is impossible without a telescope. This is completely wrong. Many experienced amateurs, who contribute scientifically valuable observations, use nothing more than binoculars. Most small telescopes that are advertised for astronomical use are worthless for looking at anything – except perhaps the Moon – so resist the temptation to rush out and buy one. If you have generous relations try to dissuade them from buying you such a telescope as a present.

It is far better to start with a pair of binoculars, but there are a number of factors to take into account when making a choice. Binoculars are specified by two figures, the magnification and the aperture, usually shown as '8 x 40', '7 x 50', or some similar combination. The first of these is the magnification, and the second is the aperture (or clear diameter), in millimetres, of the **objective** lens, i.e., the lens through which light enters the instrument.

In astronomy, the larger the aperture, the better, but if the binoculars are to be hand-held (as most are), anything larger than about 50 mm will be too heavy to keep steady for any length of time. Most larger binoculars need to be held on some form of mounting and are, in any case, much more expensive. Similarly, magnifications of more than 8 are unsuitable for hand-held binoculars. (Zoom binoculars are not suitable for astronomical use.)

Even if the magnification is not given, it is easy to determine. Hold the binoculars up, away from your eyes, in front of a plain illuminated surface (such as the sky). You will see a small illuminated circle in each eyepiece. This is called the **exit pupil**. Measure its diameter by holding a graduated scale against the eyepiece. Divide the diameter of the objective by that of the exit pupil, and you have the magnification. (The same applies to a telescope.) For astronomical use, the diameter of the exit pupil should not exceed 7 mm, which is about what one obtains with 7 x 50 binoculars.

The exit pupil should also be perfectly circular, with no flat sides which, if present, indicate that the internal prisms are cutting off some of the light. This problem is quite common in some (but not all) cheap and medium-priced binoculars.

For preference, all the optical surfaces, both external and internal, should be anti-reflection coated. This coating usually has a purple or amber tint. Expensive binoculars are coated throughout, but frequently cheap instruments are coated on just the outer surfaces. Check by holding the binoculars under a strip light and moving them until you can see the small reflected images of the light in the various internal optical surfaces. A white image shows that the surface in uncoated. Check both the objective and eyepiece ends. You may have to compromise on this point, because fully coated binoculars cost more.

Check that you are able to obtain really sharp focus with each side in turn, and both together. Use a really distant object for this test. The best binoculars have individual focusing on both eyepieces, but this is uncommon on most pairs. If you wear spectacles, make sure that you can obtain suitable correction when you take them off.

Try to find a distant object that shows a sharp line with a strong black and white contrast (such as the shadowed side of a building against the bright sky). There should be no sign of any coloured fringe. This problem is more commonly found in old binoculars.

Finally, make the most important test. Find a distant object with an even, sharply defined horizontal line, such as the top of a roof. Focus on the object and then slowly move the binoculars away from your eyes, until you can see two separate images. In good binoculars the two images of the roof will be aligned horizontally, as shown in the diagram (overleaf). In many binoculars one side will appear slightly higher than the other, as shown in the second diagram. If they are parallel, and not too far apart, your eyes will actually compensate for this error. What you do not want is for one image to be rotated relative to the other, as shown in the third diagram. Quite unconsciously, your brain tries to rotate your eyes in an effort to compensate, and this leads to considerable eye-strain. This fault is the worst that you may encounter and is to be avoided if at all possible.

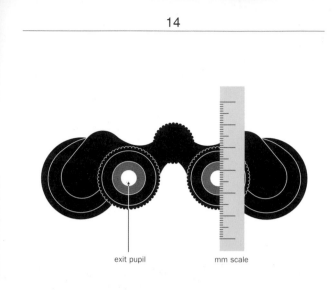

exit pupil   mm scale

Measuring the exit pupil of binoculars

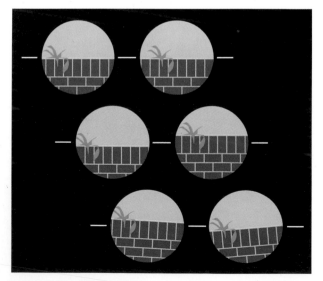

Testing misalignment of binoculars

Using binoculars

The image given by any binoculars improves greatly if you are able to hold them perfectly steady. (The improvement offered by the expensive, image-stabilized binoculars that are now on the market is quite dramatic.) Try to avoid bending the neck too far backwards. If you use an adjustable garden chair steady your elbows on the arms. When standing, bracing yourself against a wall, or resting your elbows on something will also help. You may also like to try fastening a light chain or stout cord to the centre of the binoculars. If you stand on one end and bear up gently on the other, you will find that the image improves.

Under cold conditions you may experience trouble with dew that forms on the objectives and on the eyepieces. It is easy to make dewcaps for the objectives out of matt black card to project 5–6 cm beyond the lenses, like lens hoods. Eyepiece dewing is more difficult, because the lenses are exposed to warm moist air from your eyes. All you can really do is fan them with a piece of card to clear the dew. If the lenses are dewed at the end of an observing session, make sure the dew has evaporated before fitting the lens caps and storing the binoculars away.

Treat binoculars with care. Always use the neck strap, and avoid knocks if possible. The design of most cheap and medium-priced binoculars is such that their prisms may become misaligned when they are subjected to shocks, leading to the optical errors that we have just described. Roof-prism binoculars are more robust, but are far more expensive.

Finally, although we have discussed binoculars, old-fashioned opera glasses have their uses. Older ones do not have all the modern refinements, such as anti-reflection coating, but their low magnifications of perhaps 3 or 4 give excellent wide-field views of the Milky Way and other parts of the sky.

PHOTOGRAPHING THE NIGHT SKY

It is quite easy to take photographs of the night sky. Some people find it difficult to relate charts to what they see in the sky, and find it easier to learn the constellations from photographs, which bear a close resemblance to the actual appearance of the sky.

You can use more or less any camera, although it does need to have a 'B' setting, to allow you to take manual exposures of several seconds. One problem is that many modern cameras no longer have a mechanical setting, and rely upon battery power for powering the shutter and, in single-lens reflex cameras, for raising the mirror. Long exposures may drain batteries very quickly. You may need to check this point with the camera's manufacturer.

Apart from a tripod, you need a cable release, preferably of the locking type. Because vibration is to be avoided at all costs,

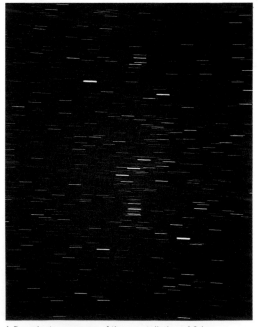

A five-minute exposure of the constellation of Orion

some photographers do not use the camera's shutter to control the exposure, but instead hold a black card or other object in front of the lens (a black hat is ideal) before opening and locking the shutter. The exposure is given by removing and then replacing the card (or hat) after the correct amount of time.

The standard 50-mm lenses on most cameras using 35-mm film cover about 35 x 47° of sky, which is a nice size for some of the most interesting constellations, such as Orion. At full aperture and using 400 ISO film, try exposures of 10, 15, 20, and 30 seconds. The stars will appear as short trails. If you have dark skies you can give much longer exposures to create striking photographs with long trails, although most constellations become more difficult to recognize in such photographs. Favourite areas are Orion and Ursa Minor, which show straight and circular trails respectively.

Always use a lens hood, which not only prevents stray light from reaching the lens, but also helps to prevent dewing. A skylight filter is also advisable, even though it does theoretically cause a slight light-loss, because it will protect the actual lens from dew and other potential damage.

Choice of film is a little difficult. Print (negative) film may be convenient, but standard processing and printing techniques tend to emphasize the background sky at the expense of the stars. Transparency (positive) film is best, and the Ektachrome series is particularly good at producing black (rather than greenish) backgrounds. Always ask for any astronomical films to be returned to you uncut, because the processing machinery often has difficulty in locating the edges of dark images, and there is a risk that it may cut through the centre of each picture.

Sighting on the required area may pose problems. With some cameras the image through the viewfinder is too faint for you to see any stars. You may need to make up a simple peep sight to indicate the area that will be included.

Don't be disappointed if the image of the Moon turns out to be extremely small. With a 50-mm lens it is about one hundredth of the width of a full frame. You need a much longer lens to obtain a bigger image, and then you also need to drive the camera mount to track the Moon. That is something you might like to try at a later stage.

THE CELESTIAL SPHERE

It helps to know a few of the special terms for parts of the sky. All the stars and other objects seem to lie on a vast sphere, known as the **celestial sphere**. As seen by an observer, at any one time half of the celestial sphere lies below the horizon. The point directly overhead is the **zenith**, and that beneath one's feet is the **nadir**. The line running from due north on the horizon, up through the zenith and down to the south is the **meridian**. All celestial objects are at their highest **altitude** as they cross the meridian. The position of an object may be described by its altitude above the horizon, measured in degrees, and by its **azimuth**. The latter is measured around the horizon from north (0°), through east (90°), south (180°) and west (270°), to the point where a vertical line through the object meets the horizon.

The celestial sphere appears to rotate from east to west around an axis running between the **North** and **South Celestial Poles**, which coincides with the Earth's rotational

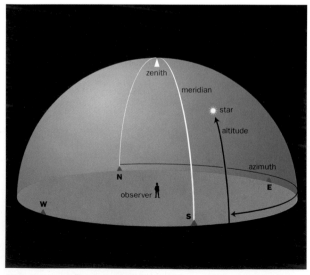

Important terms for parts of the sky seen by an observer

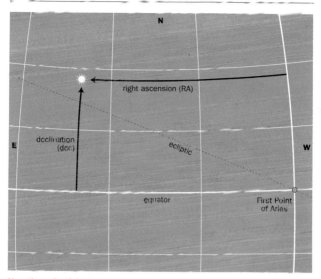

How the celestial co-ordinates of a star are measured

axis. As on Earth, the **celestial equator** divides the sky into northern and southern hemispheres.

Although it is not used to any great extent in this book, it helps to know how astronomers specify the exact position of objects in the sky. They use a system of **declination** and **right ascension**, which are similar to latitude and longitude. Declination, like latitude, is measured in degrees north or south of the (celestial) equator. Right ascension, unlike longitude, is not measured in degrees. Instead, like the day, it is divided into 24 hours, each consisting of 60 minutes, each of which is further sub-divided into 60 seconds. It also increases towards the east, so that as time passes, the right ascension on the meridian (i.e., due South) also increases.

Although the lines of right ascension and declination are shown on the individual constellation charts at the end of this book, they are not used on the introductory or monthly charts.

THE VISIBLE REGION OF THE SKY

The constellations that you can see during the course of a night or at some time during the year depend on your latitude. At the equator you would be able to see every star in the sky at some time of the year. By contrast, at the North or South Pole, half of the sky is always above the horizon, and the other half is permanently invisible.

Anyone watching the northern sky throughout the night would find that some stars and constellations remain visible all the time, even though their apparent positions change as the Earth rotates. These **circumpolar stars** are always above the horizon (even in daylight). They appear to circle a fixed spot in the sky, the North Celestial Pole (p.18). Observers in the northern hemisphere are fortunate, because a bright star, called Polaris, lies very close to this point.

Long-exposure photographs show clearly how circumpolar stars circle this point in the sky once a day. Even Polaris itself leaves a tiny trail on photographs, because it is not precisely at the North Celestial Pole. (There is no conspicuous star near

A satellite trail crossing Ursa Minor, with Polaris on the left

A one-hour exposure of the northern circumpolar stars

the South Celestial Pole, so in the southern hemisphere the stars appear to be circling a large empty patch of sky.)

The size of the northern circumpolar region increases the closer you are to the North Pole. Because circumpolar constellations are visible on any clear night, they are the first ones to learn. They are described on pp.30–35. Once you can recognise these they will guide you to the other constellations that are visible for only part of the night or at particular times of the year.

Stars just outside the circumpolar region dip below the northern horizon for a short period during the night, so they are visible on most nights. Stars near the celestial equator are above the horizon for about six hours at a time, and are lost in daylight for roughly six months of the year. South of the celestial equator, the period of visibility shortens the closer the stars are to the South Celestial Pole, with the southernmost circumpolar stars permanently invisible below the horizon.

THE CHANGING SKY

The appearance of the sky changes throughout the night, from day to day, and over the course of a year. This sometimes causes confusion in people who casually look at the sky from time to time. Once you understand the reasons for these changes, however, they cease to be a source of difficulty.

There are three main causes for the variations. The simplest to understand is the rotation of the Earth. Strangely, although nearly everyone knows that the Sun seems to cross the sky not because it is moving but because the Earth rotates on its axis, many people fail to realise that this causes the stars to behave in just the same way. Looking south just after sunset, for example, you see completely different stars from those you see just before sunrise.

The other causes of alterations in the appearance of the sky are described later. They are related to the seasons (p.24), and to the movements of the Moon (p.25) and planets (p.25–27).

Our 24-hour civil day (or **solar day**) is based upon the apparent motion of the Sun. It is the average interval from noon to noon, when the Sun crosses the meridian. In fact, relative to the distant stars, the Earth's true rotation period (a

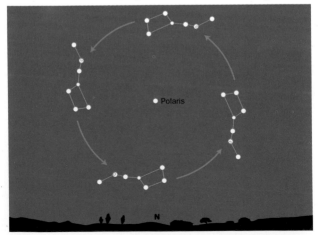

The Plough at 22:00 hours, in winter (right), spring, summer and autumn

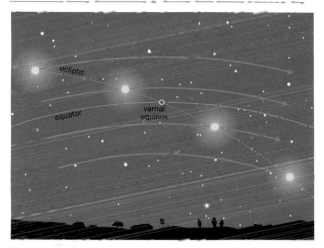

The changing path of the Sun around the spring equinox

sidereal day) is slightly less (about 23^h 56^m 4^s). This difference of nearly four minutes arises because, during the course of a day, the Earth moves along its orbit, so it must rotate a little farther for the Sun to reach the same position in the sky.

This means that relative to the Sun (and our clocks), and quite apart from their daily rotation, the constellations slowly move across the sky from east to west as the days go by. Constellations that have been in the night sky for months sink beneath the western horizon and are no longer visible after sunset. Similarly, new constellations first become visible in the east just before sunrise and slowly appear earlier and earlier in the night, until they too eventually disappear in evening twilight. The whole cycle takes a year to complete.

MOTIONS OF THE SUN, MOON & PLANETS

Everyone knows that the Sun's apparent path against the sky (the **ecliptic**) lies higher in summer than in winter. This is because the Earth's axis is not at right angles to its orbit, but tilted by about 23.4°. (If it were at right angles, the ecliptic and the celestial equator would coincide and we would have no seasons.) The Sun reaches its highest point at the summer **solstice** (on June 20 or 21) and its lowest at the winter solstice (on December 21 or 22). The slight variations in date arise because there are not an exact number of days in our calendar year.

As the Sun appears to move along the ecliptic, it crosses the celestial equator twice: from south to north at the vernal (spring) **equinox** (on March 20 or 21), and from north to south at the autumnal equinox (on September 22 or 23). The vernal equinox is particularly important, because it is used as the zero point from which right ascension (p.19) is measured. This point is also known as the **First Point of Aries**, and indicated by the sign ♈.

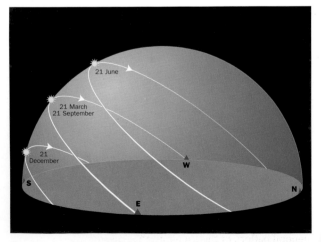

The changing altitude of the Sun throughout the year

The Sun rises towards the north-east in summer and towards the south-east in winter. At the equinoxes, it rises due east and sets due west. In this it is unlike any star, which always rises at the same point on the horizon.

Movement of the Moon

Because it is orbiting the Earth, the Moon slowly moves from west to east by slightly more than its own diameter in an hour. Relative to the stars, it takes 27.32 days to complete an orbit, but because of the difference between sidereal and solar days, mentioned earlier (p.22), it requires about two days more to return to the same position relative to the Sun, from New Moon to New Moon, for example.

Anyone watching the Moon over several months would realise that it behaves like the Sun: it is sometimes higher or lower in the sky. What is not obvious, however, is that the Moon's motion is far more complex. Its orbit around the Earth is not only inclined to the ecliptic, but also moves in space. As a result the Moon follows a complicated path that takes it varying distances above and below the ecliptic. It may at times be low on the horizon, and at others high overhead. Its rising and setting points vary accordingly. Luckily, the Moon is normally easy to see, so we need not discuss these changes in detail.

The Zodiac

The band of sky where the Moon may be found, approximately 8° on either side of the ecliptic, is known as the **zodiac**. This originally consisted of 12 constellations. With the changes in constellation boundaries introduced over the centuries, and because of the motion of the Earth's axis (an effect known as **precession**, which we need not discuss here) parts of other constellations now form part of the zodiacal region, charts of which are shown on pp.132–133.

The motions of the planets

The movement of the planets is even more complicated than that of the Moon, but there is no need to discuss the reasons in great detail. It helps to have some idea of why planets are easier to observe at certain times than at others, and also the general pattern of their motions. Because the movements are

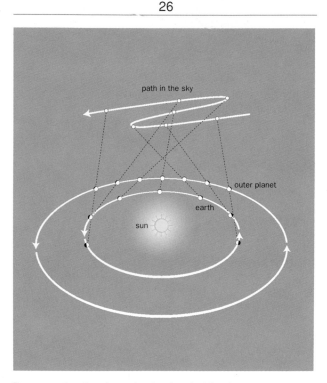

The apparent motion of an outer planet against the sky

complex, charts showing their positions for a few future years are included on pp.134–135.

Those planets that are visible to the naked eye or with small telescopes always lie within the zodiac. Their motion is generally from west to east, but occasionally the combination of motions of the Earth and a particular planet make it appear to move backwards in the sky. This movement from east to west is known as **retrograde** motion. The differing inclinations of the planetary orbits also cause a movement in declination and when these effects are combined, planets may seem to trace out loops or 'S-' or 'Z-shaped' paths on the sky.

The **outer planets**, whose orbits are outside that of the Earth: Mars, Jupiter, Saturn, and Uranus (as well as Neptune and Pluto) sometimes lie opposite the Sun in the sky. They are then at **opposition**, and are most favourably placed for observation. (They are also in the middle of their period of retrograde motion.) At times, however, these planets may appear to pass behind the Sun (and are invisible in its glare). They are then said to be at **conjunction**.

The two **inner planets**, Mercury and Venus, lie closer to the Sun than the Earth, and their paths are a complicated series of loops backwards and forwards across the sky. Obviously they are never at opposition, but they may be at conjunction, either **inferior conjunction**, when they are between the Earth and the Sun, or **superior conjunction**, when they are behind the Sun. These two planets are always fairly close to the Sun and are easiest to see when they are farthest from it either on the eastern or western side: at eastern or western **elongation**.

If you ever have trouble recognising a zodiacal constellation, it may be because a bright planet (particularly Venus, Jupiter, Mars, or Saturn) is in the region, completely changing the apparent pattern of bright 'stars'. The charts given later should help to prevent this confusion.

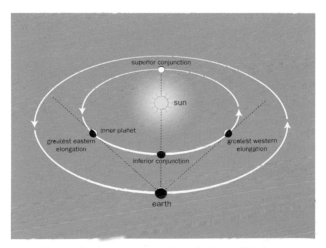

The movement of an inner planet relative to the Sun and Earth

THE NAMES OF STARS AND OTHER OBJECTS

As mentioned earlier, we use the standard, Latin names for the constellations in this book. Naming individual stars is rather more complicated. Nearly all the brightest stars have proper names, some of which are Latin (Polaris, for example), but many of which are Westernized versions of Arabic names (such as Betelgeuse). In general, these are the only names shown on the introductory and monthly charts.

In 1603, Johannes Bayer introduced a different way of identifying stars, by using a Greek letter followed by the genitive of the Latin name of the constellation. So we have, for example, α Ursae Majoris ('alpha of Ursa Major'). He lettered the stars more or less in order of brightness, so that the brightest was (generally) 'alpha' (α), the second brightest 'beta' (β), and so on. (In a few constellations he ran out of Greek letters, so he used Roman letters as well.) These names are used to this day by astronomers, so the full Greek alphabet is given here. Astronomers often use a standard set of three-letter abbreviations to indicate the constellations and these, together with the genitives, are shown later on pp.168–251.

Some of the fainter stars on the charts are identified by numbers. These are **Flamsteed numbers**, introduced by John Flamsteed in 1725, who numbered the stars in each constellation in order of increasing right ascension (p.19), i.e., from west to east.

Certain non-stellar objects, such as star clusters and galaxies, are also identified by numbers. Those beginning with the letter 'M' are objects from a famous catalogue prepared by the French astronomer Charles Messier in 1771–1781, of objects that might be confused with comets. Other numbered, non-stellar objects come from the New General Catalogue (NGC), by J.L. Dreyer, published in 1888.

A few non-stellar objects are bright enough to have been given names in antiquity, in particular the clusters known as the Pleiades (p.244), the Hyades (p.244), and Praesepe (p.182). Some of the more modern names used for particular objects will be mentioned when discussing the individual constellations.

Stellar magnitudes

The brightness of a star or planet is described as its **magnitude**. The scale originated in ancient times when the brightest objects were said to be of first magnitude; fainter ones of second magnitude, etc. There is now a strict mathematical relationship that describes the magnitude scale, but for our purposes it is sufficient to know that the faintest stars visible to the naked eye under good skies are about magnitude 6. (This naked-eye **limiting magnitude** is actually one-hundredth of the brightness of a first-magnitude star.) The scale extends beyond 1 at the brighter end. Vega (Lyrae), for example is magnitude 0, and Sirius, the brightest star in the sky, has a negative magnitude -1.4. Venus, the brightest planet, may sometimes reach magnitude -4, and the Full Moon magnitude -13.

On our charts, the limiting magnitude is about magnitude 4 for the monthly and finder charts, and 5.5 for the individual constellation charts later in the book.

Greek Alphabet

α	alpha	ζ	zeta	λ	lambda	π	pi	φ	phi
β	beta	η	eta	μ	mu	ρ	rho	χ	chi
γ	gamma	ϑ	theta	ν	nu	σ	sigma	ψ	psi
δ	delta	ι	iota	ξ	xi	τ	tau	ω	omega
ε	epsilon	κ	kappa	ο	omicron	υ	upsilon		

THE CIRCUMPOLAR STARS

The best way of locating any object, whether a constellation, a particular star, or some other object, is by the method known as 'star-hopping': using a pattern of known stars to guide you to unfamiliar ones. This may sound crude, or even simplistic, but don't be fooled. An identical procedure is used by many advanced amateur astronomers when locating faint objects, even with powerful telescopes, because it is often simpler and faster than relying on more sophisticated means. The human eye and brain are extremely clever at detecting and remembering simple patterns, so the method is actually very effective.

Because they are visible on any clear night it is sensible to start with the northern circumpolar constellations. There are five important ones: Ursa Major, Ursa Minor, Cassiopeia, Cepheus, and Draco. These are shown on the accompanying chart, and you will see that, in theory, parts of several other constellations are also circumpolar.

The altitude of the North Celestial Pole is the same as the observer's latitude, so for 50°N, for which the charts in this book are drawn, the altitude of the pole is 50° and a circle of

The Plough: this is where you start

Constellations in the northern circumpolar region

sky 100° across is circumpolar. In practice, however, the region close to the horizon – even if it is clear of physical obstacles – is often obscured by haze (and light pollution). In addition, the closer stars are to the horizon, the more light is absorbed. Only a few bright stars may be seen within about 10° of the horizon. Capella, α Aurigae, is about 43° from the North Celestial Pole and, because it is bright, is generally visible just above the horizon when due north.

An effect known as **refraction** helps to counteract the loss of visibility close to the horizon. The paths of light through the atmosphere are curved and this 'lifts' the images of stars (and other objects) so that they are visible for longer or at lower altitudes. When the Sun, for example, appears to set, it is actually about 34' (just over half a degree) below the horizon. The images of stars, the Moon, and the planets are raised by a similar amount.

Finding Polaris and Ursa Minor from the Plough

Begin learning the sky by finding the seven stars of the Plough, which forms part of the constellation of **Ursa Major** (the Great Bear). Their pattern is so distinctive that nearly everyone knows them already, but if you are uncertain, you will be able to recognise them from the charts shown here. The changing orientation in the sky throughout the year is shown on the monthly charts. The two stars at the end of the 'bowl' are α and β Ursae Majoris, otherwise called **Dubhe** and **Merak**, respectively, and together known as the **Pointers**.

A line from Merak to Dubhe, extended about five times the distance between them, points to an isolated bright star. This is **Polaris**, α Ursae Minoris, the Pole Star (p.20). It is the brightest star in the small constellation of **Ursa Minor** (the Little Bear), which consists mainly of a rough rectangle of stars, and three others forming the 'tail', with Polaris at the tip. The two brightest stars in the rectangle, nearest the Plough, are β and γ Ursae Minoris, **Kochab** and **Pherkad**, respectively, also known as the **Guards**

On the opposite side of the Pole to the Plough is the distinctive constellation of **Cassiopeia**, with a conspicuous group of five stars that form the letter 'W' (or 'M' at different times of night or of the year). Regardless of when you are observing, either Ursa Major or Cassiopeia will be readily visible, and help you to orient yourself on the sky, even if a large part of the northern sky is obscured by trees or buildings.

A line from Ursae Majoris, the first star in the handle of the Plough (or tail of the bear, if you wish) through Polaris, extended by about the same distance, leads to γ Cassiopeiae (also known as **Cih**), the central star of the five that form the 'W'. Cassiopeia actually lies in a fairly dense region of the Milky Way, so a large number of other stars are also visible in this area on a clear night.

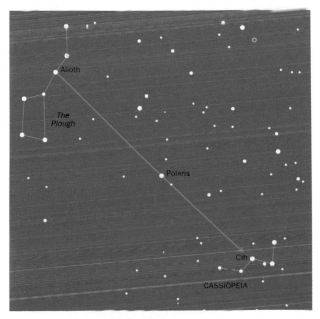

Finding Cassiopeia from the Plough and Polaris

The other two constellations of immediate interest in the northern circumpolar region are rather fainter. The constellation of **Cepheus** has five main stars that form a shape usually described as like the slightly lop-sided gable-end of a house. The line from the Pointers in the Plough through Polaris, extended by about half that distance, points roughly to the centre of the triangular 'gable'. The star at the apex, γ Cephei (**Errai**) lies close to the line between Polaris and β Cassiopeiae (**Caph**). A line through α and β Cassiopeiae points to the brightest star in Cepheus, α (**Alderamin**) at one corner of the base. Like Cassiopeia, Cepheus includes part of the Milky Way, so there are large numbers of fainter stars in the lower part of the constellation.

Draco (the Dragon) is such a long, winding constellation that it is a little difficult to recognize at first. Find the two stars γ and δ Ursae Majoris (**Phecda** and **Megrez**) at the opposite end of the body of the Plough from the Pointers. Extend a line

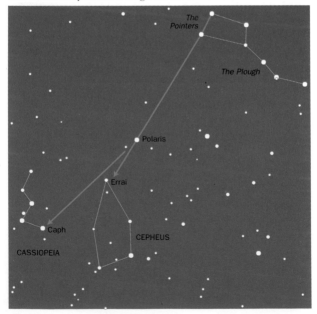

Locating the constellation of Cepheus

Finding the head of Draco

from them across the sky south of the Guards in Ursa Minor, for about eight times their separation. This takes you to a distinctive 'lozenge' of four moderately bright stars, the brightest of which, γ (**Eltanin**) is farthest from the pole. These form the 'head' of Draco. From there, the body runs first towards the side of Cepheus, then turns back and follows a winding course above the Plough. It eventually ends between Ursa Major and Ursa Minor at a star just above the Pointers.

There is one other constellation, Camelopardalis, that is fully circumpolar, but it is rather faint, and will be described later (p.180). The same applies to several other constellations, parts of which are also circumpolar.

USING THE MONTHLY CHARTS

The circumpolar constellations may be learned at any time, but most others are easiest to see at particular times of the year. The following pages give a set of charts for each month.

Two charts show the overall appearance of the sky, one looking north, and one looking south. They are drawn for a latitude of 50°N, which is a compromise suitable for most of Europe. Observers farther north will see more stars above the northern horizon and lose a few in the south. The opposite will apply for observers farther south, who will gain stars in the south and lose some in the north. The charts have also been drawn so that they show the area around the zenith, directly overhead, without the distortion sometimes encountered. They have a generous overlap, so the whole of the visible hemisphere is covered.

The date and time for which each pair of charts is drawn warrants a little explanation. They show the sky as it appears in the middle of the month, at 22:00 local standard time (10:00 p.m.) during the winter months, and 23:00 local summer time (11:00 p.m.) when daylight saving time is in operation. This time was chosen because it is then reasonably dark, even in summer, although observers in the far north will never lose the twilight glow on the northern horizon. For Britain, the local times are GMT during the winter, and BST in the summer. Summer Time normally begins on the last Sunday of March, and ends on the last Sunday of October, but slight differences may be adopted in certain years.

The charts may also be used at other times of the night. For example, the first pair of charts are drawn for 22:00 on January 15, but they are also suitable for 23:00 on January 1, and 21:00 on February 1. These dates and times are listed for each pair of charts, so you can easily find out what stars are visible on any particular night. There is a difference of two hours per month, so if you want to see how the sky appears two hours earlier in the night – provided it was dark, of course – turn back a month; two hours later, turn to the next month.

There are two or three other charts for each month. These show you how to find specific constellations, either from the circumpolar constellations that you already know, or from easily recognizable, equatorial constellations. One of the twelve main zodiacal constellations is introduced each month, together with two (or occasionally three) other constellations that are well-placed above the horizon.

Don't be alarmed by the thought of learning all these different stars and constellations. You will soon find that you can identify the more prominent ones, and you can 'fill in the gaps' with the fainter, smaller constellations when it is convenient. The table shows the pages with the finder charts for individual constellations.

To prevent confusion, the monthly charts show only the brightest stars in each constellation. (Fainter stars are included on the individual constellation charts later in the book.) If a meteor shower (p.160) is active during the month, this is also mentioned.

Finder charts for the constellations below may be found on the following pages:

The key features below are not illustrated with finder charts but are described on the following pages:

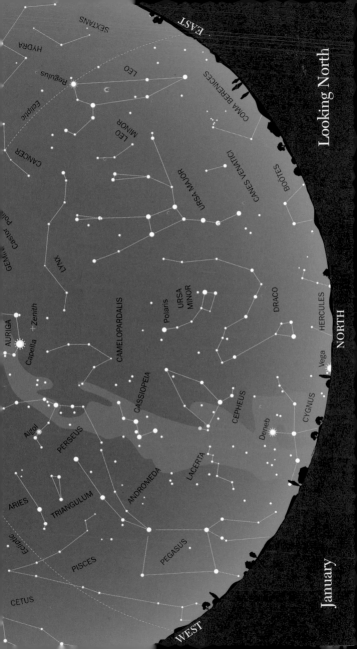

The whole of the constellation of Ursa Major

JANUARY LOOKING NORTH

JANUARY CHARTS CAN BE
USED AT THE FOLLOWING
DATES AND TIMES:

Jan.01: 23:00 GMT
Jan.15: 22:00 GMT
Feb.01: 21:00 GMT

Most of the circumpolar constellations are easy to see in the northern sky at this time of year. Ursa Major stands vertically in the north-east, and the body of Ursa Minor is below the pole in the north. The head of Draco is low on the northern horizon, and may be difficult to see under some conditions. Both Cepheus and Cassiopeia are readily visible in the north-west, and even the constellation of Camelopardalis is high enough in the sky for its faint stars to be readily visible. The Great Square of Pegasus (p.95) is setting in the west, while Leo (p.59) is rising in the east.

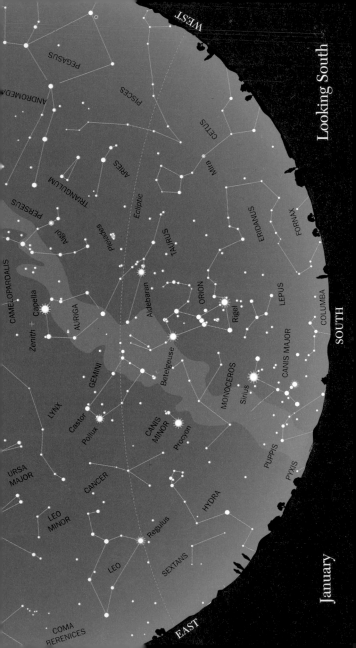

Orion, key constellation in the winter skies

JANUARY LOOKING SOUTH

In the south, the sky is dominated by **Orion**. This is the key constellation during the winter months, when it is visible at some time during the night. Its highly distinctive shape, with the line of three stars that form the 'Belt', is unmistakable. Depending on your eyesight, **Betelgeuse**, the bright star at the north-western corner, may have a reddish tinge. **Rigel**, on the opposite side of the constellation, is a brilliant bluish-white. A vertical line of three 'stars' forms the 'Sword' hanging below the Belt. Under dark, clear skies, the central 'star' appears as a hazy spot. This is actually the great Orion Nebula (p.226).

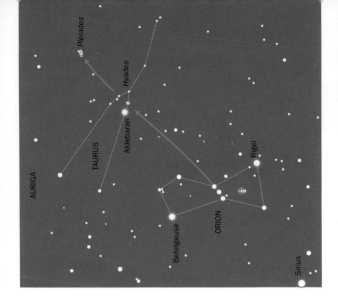

JANUARY

METEORS

JANUARY 1–6,
maximum 4:
QUADRANTID meteor
shower, bright
bluish- and yellowish-
white meteors
(see p.160).
MAXIMUM HOURLY RATE:
70

Follow the line of Orion's Belt up to the north-east to find orange-tinted **Aldebaran** in **Taurus** (the Bull). A conspicuous nearby 'V' of stars, pointing down to the south-west, is known as the **Hyades** cluster. Continuing farther along the same line from Orion takes you slightly north of a bright cluster of stars, the **Pleiades**, or Seven Sisters. Even the smallest pair of binoculars reveals this cluster as a beautiful group of bluish-white stars. The most conspicuous other stars in Taurus are two, directly above Orion, that form an elongated triangle with Aldebaran. In the past, the northernmost of these stars, β Tauri, was often regarded as part of the constellation of Auriga.

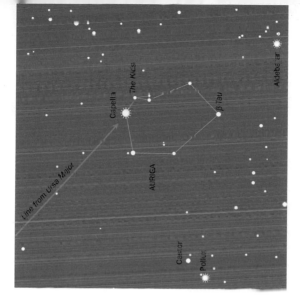

JANUARY

Almost directly overhead (i.e., at the zenith) at this time of year is brilliant **Capella**, the brightest star in **Auriga** (the Charioteer), with, slightly to its west, a triangle of fainter stars, known as 'The Kids'. (Ancient mythological representations of Auriga show him carrying two young goats.) Capella may also be found by extending to the west the line formed by the top two stars in the bowl of Ursa Major (δ and α UMa). It lies at about five times their separation. When the northern most bright star in Taurus, β Tauri, is included the main body of Auriga forms a large pentagon on the sky, with the Kids lying part-way along one side.

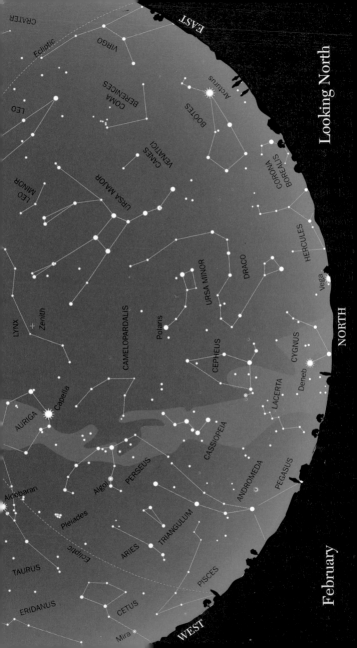

Auriga, with Capella and the triangle of the Kids (top centre).

FEBRUARY LOOKING NORTH

FEBRUARY CHARTS CAN BE USED AT THE FOLLOWING DATES AND TIMES:

Feb.01: 23:00 GMT
Feb.14: 22:00 GMT
Mar.01: 21:00 GMT

Ursa Major is now high in the eastern sky, while Cassiopeia is lower down towards the north-western horizon. The inconspicuous constellation of Lynx is at the zenith. Cepheus and the head of Draco are in the north, where, with a clear view down to the horizon, Deneb may just be seen. **Arcturus** in **Boötes** (p.51) is at about the same altitude in the north-east, but rises higher as the night goes by. Capella and Auriga are high overhead in the west, where Perseus and Taurus are clearly seen. Most of Andromeda is still visible, although **Sirrah**, the star at the corner of the Great Square of Pegasus is not easy to make out so close to the horizon.

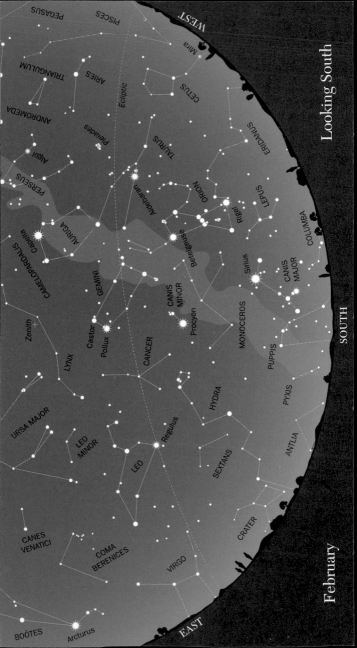

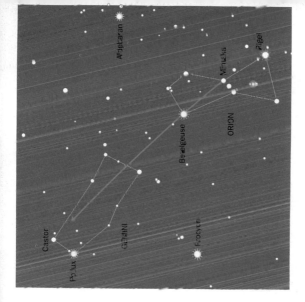

FEBRUARY LOOKING SOUTH

Orion (p.41), remains clearly visible towards the south-west. High on the meridian is the zodiacal constellation of **Gemini** (the Twins). A line from the most northerly star of Orion's belt (**Mintaka**) through Betelgeuse points to **Castor** (α Geminorum), with the slightly brighter **Pollux** (β Gem) to the south. The main part of the constellation consists of two lines of stars that run back towards Orion, forming an elongated rectangle.

FEBRUARY

The line of Orion's belt, followed down towards the south, points approximately towards **Sirius**, the brightest star in the sky, which is so conspicuous that it is easily recognized. It lies in the constellation of **Canis Major** (the Great Dog), and a chain of stars runs from it farther down towards the south-east. Two fairly bright stars lie to the south-west of this line.

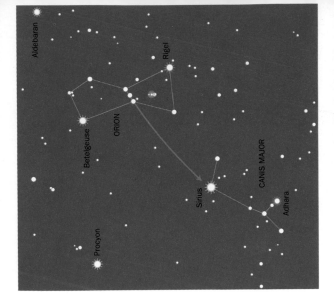

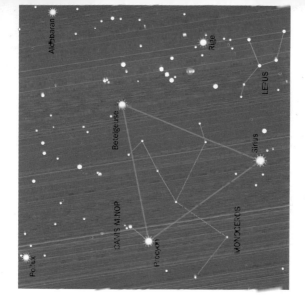

FEBRUARY

Up towards Gemini, a single bright star,
Procyon, forms an equilateral triangle with
Betelgeuse and Sirius. It lies in **Canis Minor**
(the Little Dog), an inconspicuous
constellation largely consisting of Procyon (α
Canis Minoris) and one other star to the
north-west. Between Canis Major and Canis
Minor there is the very faint constellation of
Monoceros (the Unicorn).

Lepus (the Hare), another small faint constellation, lies below Orion. Its main stars form
a three-branched shape, rather like a miniature
version of Perseus (p. 108).

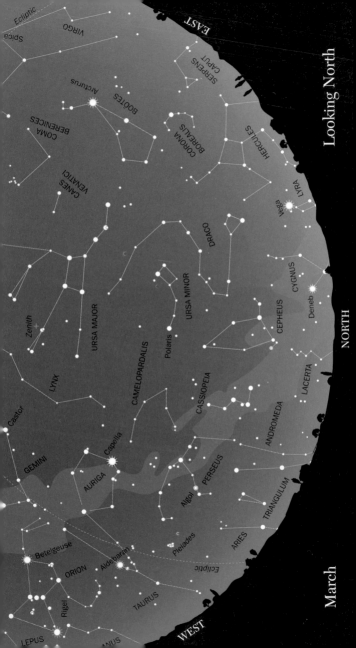

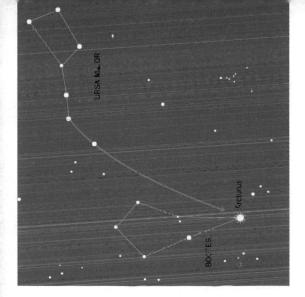

MARCH LOOKING NORTH

MARCH CHARTS CAN BE
USED AT THE FOLLOWING
DATES AND TIMES:
Mar.01: 23:00 GMT
Mar.15: 22:00 GMT
Apr.01: 21:00 GMT
    (22:00 BST)

Between north and north-east, Vega in Lyra (p.222) can be seen just above the horizon. Although **Vega** is theoretically circumpolar from 50°N, and is bright, it (like Deneb in Cygnus) is often lost in the haze or mist of the northern horizon. Ursa Major is now high overhead. If you extend the arc formed by the stars of its 'tail', you come to brilliant, yellowish **Arcturus**, fairly high in the east. This is the brightest star in the northern hemisphere, and the fourth brightest in the sky. It lies in the constellation of **Boötes** (the Herdsman), the shape of which has been likened to a kite, an ice-cream cone, or the letter 'P'. Arcturus lies at the base, farthest from the north celestial pole.

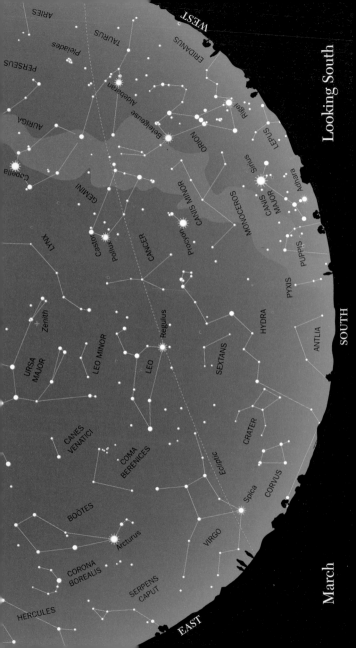

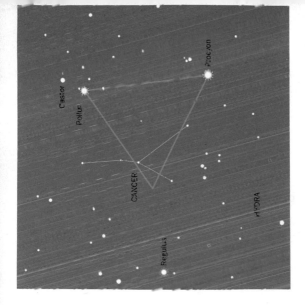

MARCH LOOKING SOUTH

Orion is now sinking in the western sky, but
most of the constellation is still visible,
although Rigel is getting rather close to the
horizon. Sirius is at about the same altitude,
but because of its brightness it is still easy to
see. Slightly to the west of the meridian is the
zodiacal constellation of **Cancer**. Although
very faint, this is an ancient grouping of stars.
It is not easy to pick out, but if you imagine an
equilateral triangle with Castor in Gemini and
Procyon in Canis Major as two of the corners,
the other corner (to the east) lies close to the
centre of Cancer. Three 'legs' extend to the
north, south-west, and south.

MARCH

Below Cancer, and slightly south of the mid-point of a line joining Procyon and Regulus (p.59), lies an extremely distinctive little group of six stars. This asterism is the head of **Hydra,** a long, sprawling constellation that winds away towards the east. It is so long that it will be another three months before its tail is near the meridian at this time of night. Its brightest star, **Alphard,** (α Hya) is currently almost due south.

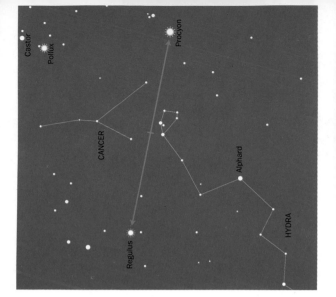

The constellation of Leo, with Regulus and the Sickle

The faint constellation of Cancer, with Praesepe (centre)

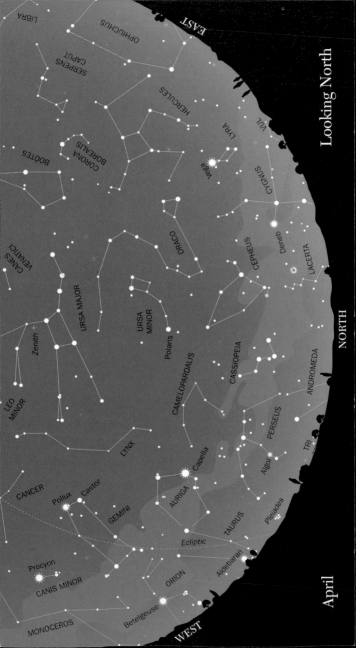

Hercules, with the 4 stars that form the 'Keystone' (centre)

APRIL LOOKING NORTH

APRIL CHARTS CAN BE
USED AT THE FOLLOWING
DATES AND TIMES:
Apr.01: 23:00 GMT
(00:00 BST)
Apr.15: 23:00 BST
May.01: 22:00 BST

Ursa Major is now right overhead, at the zenith, the most inconvenient position for observation. Vega and the other stars of Lyra are now higher in the north-east. Deneb and some of the stars in **Cygnus** are also rising into view, but the constellation remains rather too low for observation in the early part of the night. Arcturus in Boötes is extremely conspicuous in the south-eastern sky, where **Hercules** (p.67) is now easy to see. Perseus is getting low in the north-west, and Aldebaran in Taurus is skimming the horizon, which means that it is generally invisible by this time of night

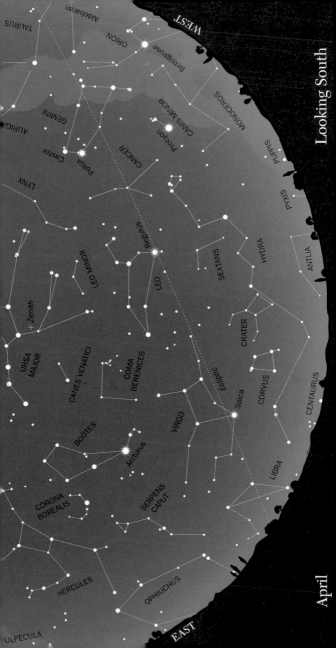

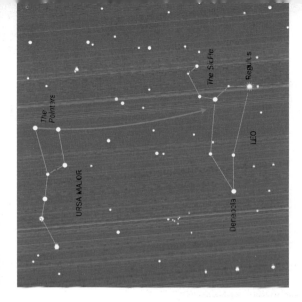

APRIL LOOKING SOUTH

The conspicuous zodiacal constellation of **Leo** (the Lion) is high in the south, slightly west of the meridian. It hardly needs any guide, but it may be found when less conveniently placed by following the line of the Pointers in the opposite direction to normal. This line takes you down not far from **Regulus**, the brightest star in the constellation. Above Regulus is the distinctive asterism of the Sickle, a backwards question-mark of stars open to the west. The second brightest star, **Denebola**, is close to the meridian at the time for which the chart is drawn.

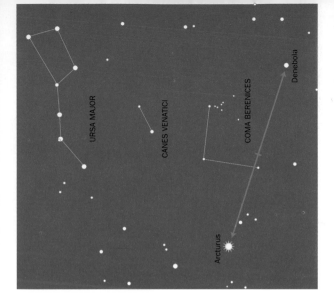

Labels on image: URSA MAJOR, CANES VENATICI, COMA BERENICES, Denebola, Arcturus

APRIL

Between Leo and Boötes lies **Coma Berenices** (Berenices Hair), a small, inconspicuous constellation, largely consisting of three faint stars arranged in a right-angle. The southernmost is roughly on the line between Denebola and Arcturus. Above Coma (as it is often called) is the tiny constellation of **Canes Venatici** (the Hunting Dogs), which effectively consists of two stars, lying beneath the curve of the tail of Ursa Major.

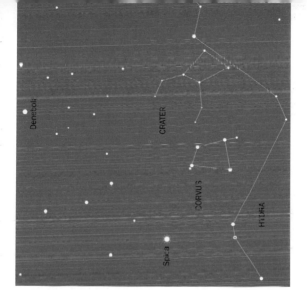

APRIL

Most of Hydra is now visible, although the southernmost stars are too low on the horizon to be seen easily. This is a good time of year to see the two small distinctively shaped constellations of **Crater** (the Cup) and the slightly brighter quadrilateral of stars that forms **Corvus** (the Crow). Both lie south of Denebola and Corvus is close to **Spica**, the brightest star in **Virgo**, which we describe next month.

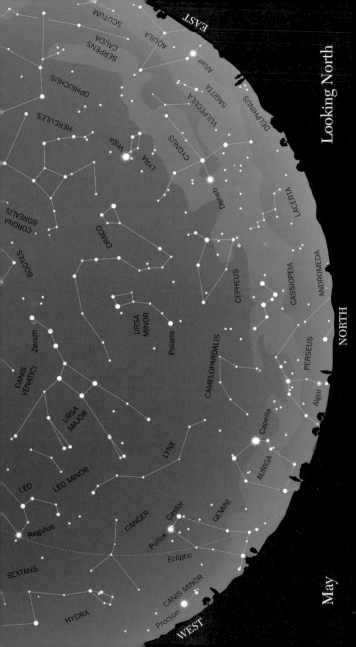

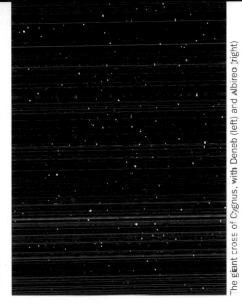

MAY LOOKING NORTH

Cassiopeia is now swinging low above the northern horizon. Ursa Major stands vertically in the west, with the end of its tail (η UMa, **Alkaid**) right at the zenith. Most of Perseus is lost in the northern twilight, but brilliant Capella is still easily visible, as are Castor and Pollux in Gemini farther to the west. In the eastern sky, Vega and the rest of Lyra are now readily visible, as is most of Cygnus. You may also be able to glimpse **Altair** in **Aquila** (p.77), and by midnight it will be sufficiently high above the eastern horizon for it to be easy to see.

The giant cross of Cygnus, with Deneb (left) and Albireo (right)

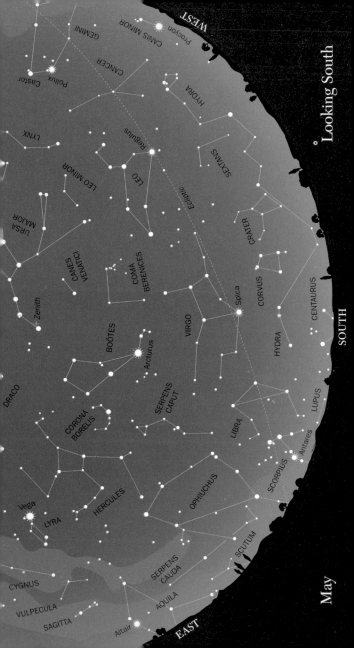

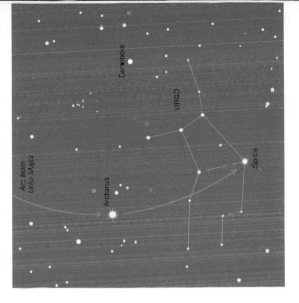

MAY LOOKING SOUTH

To the south, Arcturus in Boötes is almost on the meridian. If you extend the arc of the tail of Ursa Major, through Arcturus – as before – and onwards, you come to a conspicuous, bright, bluish-white star, just to the west of the meridian. This is **Spica**, the brightest star in **Virgo** (the Virgin), which is now best placed for viewing. Although this is the largest zodiacal constellation, Virgo does not have a very memorable shape, but Spica and four other stars form an approximately quadrilateral 'body', with 'arms' and 'legs' at each corner.

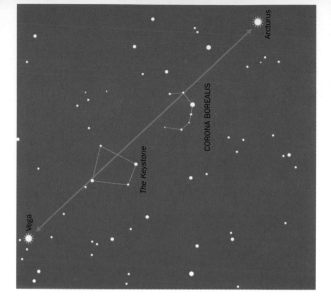

MAY

A line between Arcturus and Vega passes
through two distinct constellations. Just to the
east of Boötes, and about the same altitude as
Arcturus above the horizon, there is a readily
visible, incomplete circlet of stars, one
significantly brighter than the others. These
form the small constellation of **Corona
Borealis** (the Northern Crown).

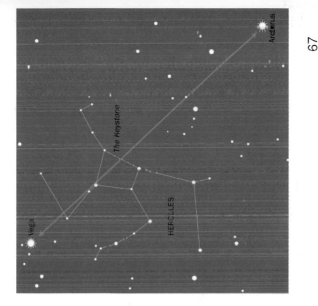

MAY

Between Corona Borea is and Vega there is a
distinctive trapezium-shaped group of four
stars. This is the asterism known as the
Keystone, which forms the 'body' of the
constellation of **Hercules**. Four 'arms' and
'legs' extend from each corner. In fact, because
of precession (p.25) **H**ercules now appears
'upside down' with his legs towards the north.

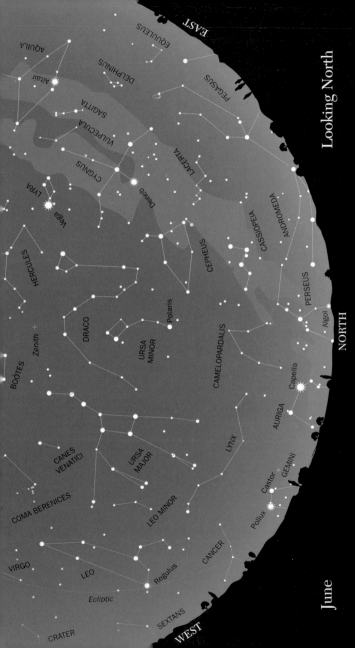

JUNE LOOKING NORTH

JUNE CHARTS CAN BE USED AT THE FOLLOWING DATES AND TIMES:

Jun.01: 00:00 BST
Jun.15: 23:00 BST
Jly.01: 22:00 BST

We have now come to the period around the summer solstice, when twilight lasts throughout the night and the sky never becomes fully dark. Some of the constellations, particularly those farthest south, may be difficult to see clearly, especially if bright moonlight also interferes.

Ursa Major is now nicely placed in the north-west, with Leo descending towards the horizon beneath it. Draco snakes its way around the sky between Polaris and the zenith. Capella is very low in the northern twilight, skirting the horizon. The brilliant **Summer Triangle** (p.77) of Vega, Altair and Deneb is very conspicuous, well above the horizon in the east.

The Summer Triangle: Vega (top), Deneb (left), and Altair (bottom right).

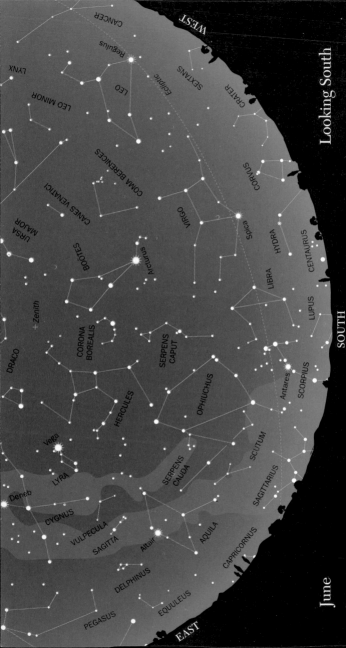

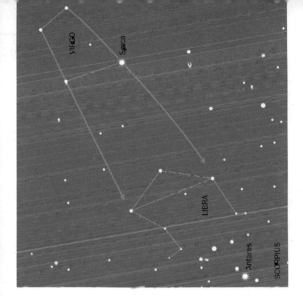

JUNE LOOKING SOUTH

In the south, the zodiacal constellation of **Libra** (the Scales) is just to the west of the meridian. This grouping is normally found by scanning eastward from Spica. A little imagination enables you to recognise the triangular beam of an old-fashioned balance, lying on its side, with two weight pans dangling from it. Although this is a very old constellation, it once formed the 'claws' of Scorpius (the Scorpion) the next zodiacal constellation to the east.

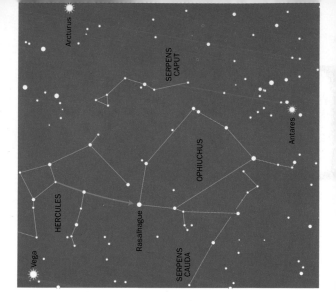

JUNE

Part of the south-eastern 'arm' of Hercules points down towards **Rasalhague** in the large constellation of **Ophiuchus** (the Serpent Bearer). This, like Cepheus (p.34), somewhat resembles the gable-end of a house, except that in this case the building is larger, with a smaller gable, and there is an additional star in the middle of the base. The southernmost part of Ophiuchus extends beyond the ecliptic, and the Sun and planets spend more time in this constellation than they do in neighbouring Scorpius (p.78).

Between Ophiuchus and Boötes lies the chain of stars of **Serpens Caput** (the Head of the Serpent). This is the only constellation that consists of two separate parts: the other half (**Serpens Cauda**) is on the eastern side of Ophiuchus.

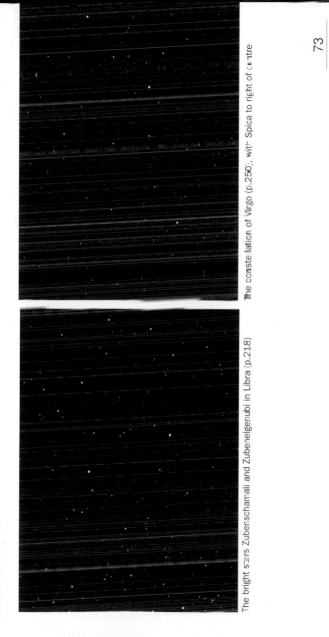

The bright stars Zubenschamali and Zubenelgenubi in Libra (p.218)

The constellation of Virgo (p.250), with Spica to right of centre

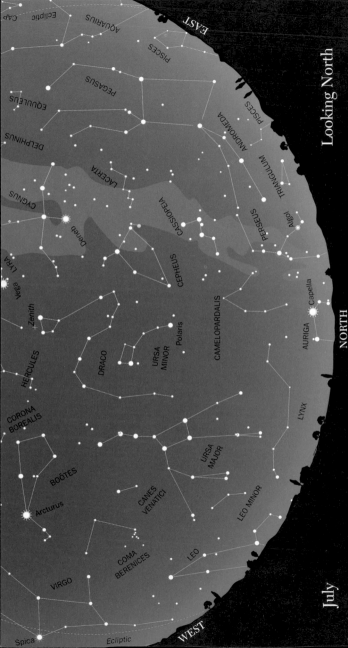

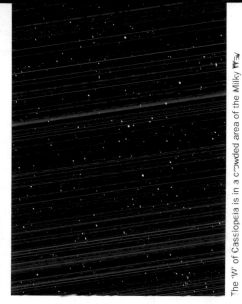

JULY LOOKING NORTH

Cassiopeia is easily distinguished in the north-east, and slightly farther to the east, the Great Square of **Pegasus** (p.95) is rising from the horizon. The three stars of the Summer Triangle, **Vega**, **Deneb** and **Altair**, are so conspicuous that they dominate the sky for several months and become extremely familiar, rather like Orion in the winter. Vega, the brightest of the three, lies almost at the zenith, north-west of a tiny parallelogram of stars. Together they make up the constellation of **Lyra** (the Lyre).

In the east, Deneb is the brightest star in **Cygnus** (the Swan). Two lines of stars form a great cross like the outstretched wings and neck of a swan, flying 'down' the broad band of the **Milky Way** towards the south.

The 'W' of Cassiopeia is in a crowded area of the Milky Way

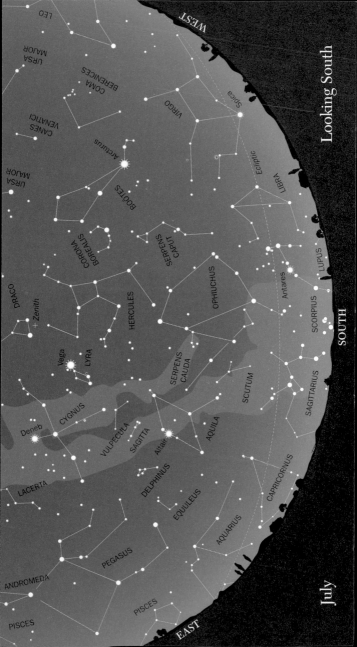

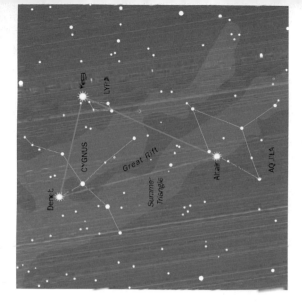

July Looking South

Dark skies are needed to see the **Milky Way** in all its glory, but under good conditions a dark bar (known as the Great Rift in Cygnus) appears to run along its centre. **Altair**, the southernmost star in the Summer Triangle, lies in **Aquila** (the Eagle) on the lower side of the Great Rift. Rather than a cross like Cygnus, the 'wings' and 'body' of Aquila form a diamond shape on the sky, with a 'neck' that also points down the Milky Way.

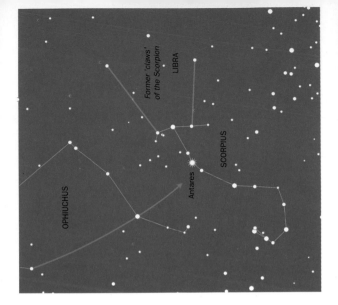

JULY

Low on the southern horizon is the deep red star **Antares** in the zodiacal constellation of **Scorpius** (the Scorpion). Its colour contrasts sharply with the yellowish Arcturus, still high in the west, and with the bluish-white and white stars of the Summer Triangle in the south-east. Three moderately bright stars lie in a north-south line between Antares and Libra, but one needs to be at lower latitudes to see the curving tail of stars that runs south of Antares and ends in the triangular 'sting'.

METEORS

Jly.15–Aug.20, maximum Jly.28 - Aquarids: double radiant, southern component more active than northern (maximum hourly rates 20 and 10 respectively). Faint meteors

The 'Teapot' of Sagittarius (p.83) is lower centre and right

Scorpius, rotated 45° anticlockwise from the chart above

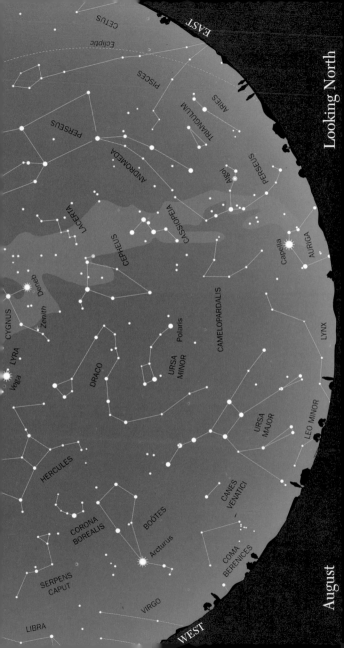

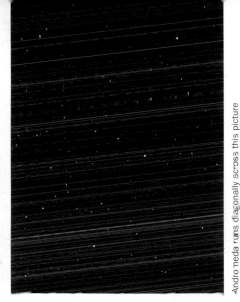

Andromeda runs diagonally across this picture

AUGUST LOOKING NORTH

AUGUST CHARTS CAN BE
USED AT THE FOLLOWING
DATES AND TIMES:
Aug.01: 00:30 BST
Aug.15: 23:30 BST
Sep.01: 22:00 BST

Boötes stands vertically in the west, and the constellations of **Corona Borealis** (p.66) and **Hercules** (p.67) are also well placed for observing. Arcturus is slowly sinking towards the horizon and will soon become difficult to see. Ursa Major stretches across the sky in the north-west, and to the north-east, Capella in Auriga and the constellation of Perseus are beginning to climb higher in the sky. A little farther round towards the south, Andromeda (p.168) and the Great Square of Pegasus are now easy to see, together with part of the constellation of **Pisces** (p.101). The Summer Triangle continues to dominate the sky overhead, and Deneb and Vega are both fairly close to the zenith.

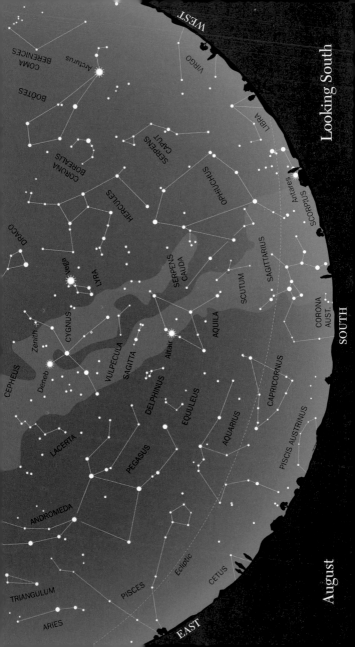

AUGUST LOOKING SOUTH

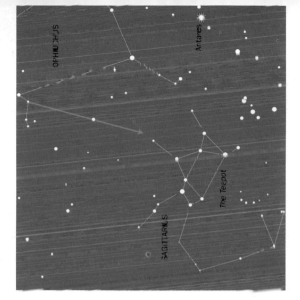

In the south-west, Ophiuchus and the two parts of Serpens are all clearly placed for observing. The eastern 'arm' of Ophiuchus points down towards the zodiacal constellation of **Sagittarius** (the Archer). This is very low or the horizon, and in fact is best seen early in the month, or even in July. None of its stars is very bright, but several of them form a distinctive pattern, known as the 'Teapot', which, once seen, is never forgotten.

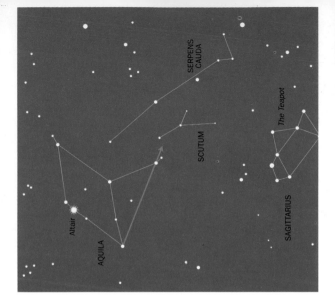

METEORS

JULY 23 TO AUGUST 20, maximum Aug.12-13:

PERSEIDS – one of the most consistent meteor showers. **MAXIMUM HOURLY RATE** about 80. Many Perseid meteors are bright and often leave trains.

AUGUST

There are several small constellations in this part of the sky, but some are difficult to recognise because their stars are not particularly bright, and they therefore tend to be lost against the multitude of stars that make up the Milky Way. Between the constellations of Serpens Cauda, Sagittarius and Aquila, lies **Scutum** (the Shield), with just four moderately bright stars, and which is probably easiest to find from the southernmost stars in Aquila.

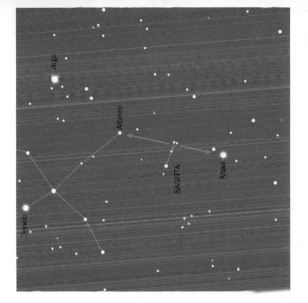

AUGUST

Farther north, between Altair and the bright star (**Albireo**) at the head of Cygnus, lies the small constellation of **Sagitta** (the Arrow), which largely consists of four stars forming a wedge shape within the Milky Way.

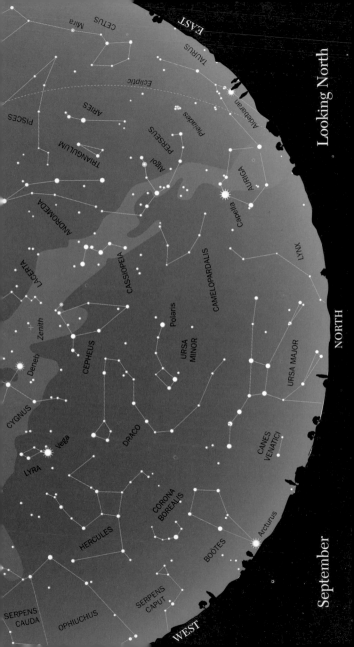

Aries (centre right) and the Pleiades in Taurus

SEPTEMBER
LOOKING NORTH

SEPTEMBER CHARTS CAN
BE USED AT THE
FOLLOWING DATES AND
TIMES:

Sep.01: 00:00 BST
Sep.15: 23:00 BST
Oct.01: 22:00 BST

In the north, Ursa Major is ow above the
horizon, but the main part of this large
constellation is readily visible. Capella, in
Auriga, is beginning to be easier to see, as it
climbs higher in the north-east. Perseus,
Triangulum, and Aries are now perfectly
distinct and the return of the beautiful
Pleiades cluster (p.42) to the right sky
indicates that autumn is upon us. It will remain
visible throughout the winter, and will soon be
joined by the magnificent constellation of
Orion, which begins to rise above the horizon
about an hour later.

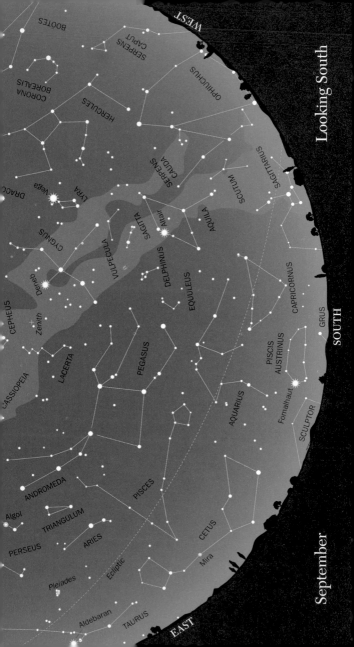

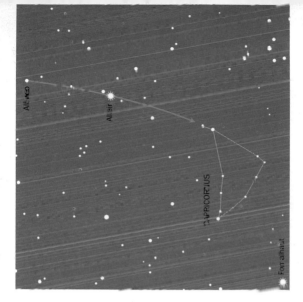

Alb reo
Altair
CAPRICORNUS
Fon alhaut

September
Looking South

The zodiacal constellation close to the
meridian this month is **Capricornus** (the Sea
Goat). It is probably best found by extending
the line from Albireo (β Cygni) to Altair by
about the same distance. This takes you to the
conspicuous group formed by α (which is
actually double) and β Capricorni. Capricornus
itself forms a large, slightly distorted triangle,
with δ Cap almost due east of β Capricorni
and about the same altitude above the horizon.

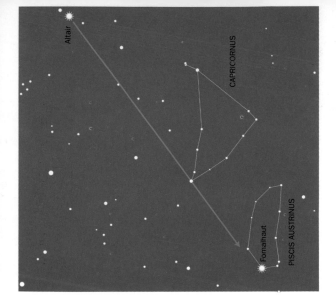

SEPTEMBER

If you have a really clear horizon to the south, under good conditions you should be able to see **Fomalhaut** in **Piscis Austrinus** (the Southern Fish). It may be found by extending the line from Altair to δ Capricorni. It is the brightest star in this part of the sky but all the others in this constellation are faint and thus difficult to see from our latitude.

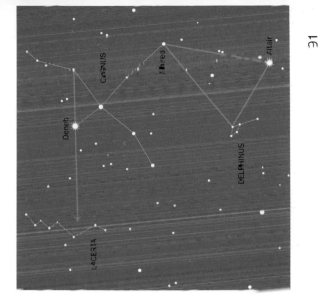

SEPTEMBER

One distinctive little constellation is
Delphinus (the Dolphin), east of Sagitta,
which was described last month. Its five stars
lie at one corner of an isosceles triangle formed
with Albireo in Cygnus and Altair in Aquila.
Not far from the zenith is another small
constellation, **Lacerta** (the Lizard), a zigzag of
stars to the east of Deneb. This is not as easy to
recognize, because it lies in a fairly crowded
part of the Milky Way.

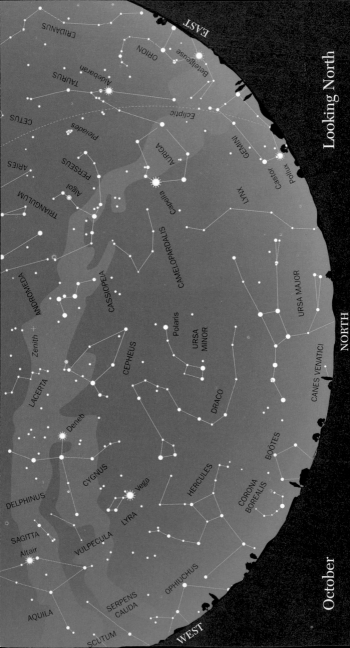

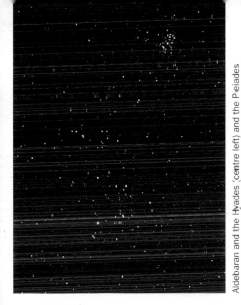

Aldebaran and the Hyades (centre left) and the Pleiades

OCTOBER LOOKING NORTH

OCTOBER CHARTS CAN BE USED AT THE FOLLOWING DATES AND TIMES:

Oct.01: 00:00 BST
Oct.15: 23:00 BST
(22:00 GMT)
Nov.01: 21:00 GMT

Capella and the Kids (p.43) are now well up in the sky in the north-east as is Aldebaran (p.42) in Taurus. Like the Pleiades cluster and Orion, these stars will become very familiar, because they are visible throughout the winter. On the opposite side of the sky, the Summer Triangle is now beginning to move down towards the western horizon. Cassiopeia, part of Cepheus, and the small constellation of **Lacerta** are high overhead in the region around the zenith. The faint circumpolar constellation of Camelopardalis is well placed for observation between Cassiopeia and Auriga in the north-east.

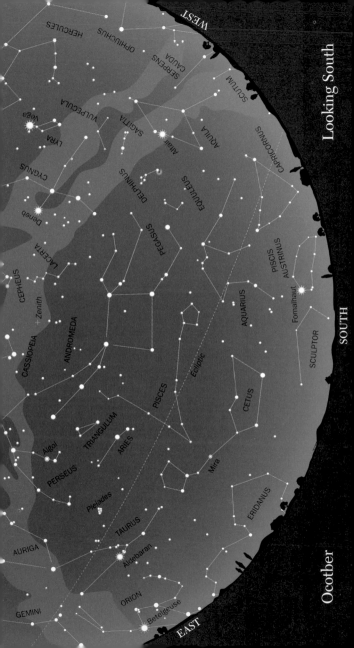

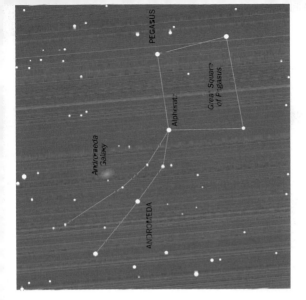

OCTOBER LOOKING SOUTH

In the south, the Great Square of **Pegasus** is right on the meridian, and quite unmistakable. In fact, the star at the north-eastern corner of the Great Square, **Alpheratz** (or **Sirrah**) belongs to **Andromeda**. Alpheratz and the three brightest of the remaining stars in this constellation form a line running eastwards and to the north. On a clear night, and provided there is not too much light pollution, most people can see the misty spot that is the distant Andromeda Galaxy (p.168).

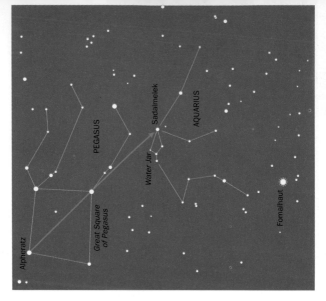

OCTOBER

The mythological representation of Pegasus, like Hercules (p.67) is upside-down on the sky, and it is not too difficult to visualize the lines of stars running west from the two stars on the other side of the Great Square as the neck and head, and two forelegs of the flying horse.

Beneath the outstretched 'head' of Pegasus is the zodiacal constellation of **Aquarius** (the Water Carrier). A diagonal line across the Great Square from Alpheratz points to the brightest star, **Sadelmelik**. Just to its east is a distinctive 'Y' of four stars. This asterism is known as the 'Water Jar' of Aquarius.

METEORS

OCT.16–26, maximum Oct.21 - ORIONIDS: fast meteors, with numerous trains. MAXIMUM HOURLY RATE about 25

The brightest 'star' is Jupiter, here in Capricornus (p.90).

Aquarius, with the 'Water Jar' (top left, centre) and the planet Saturn

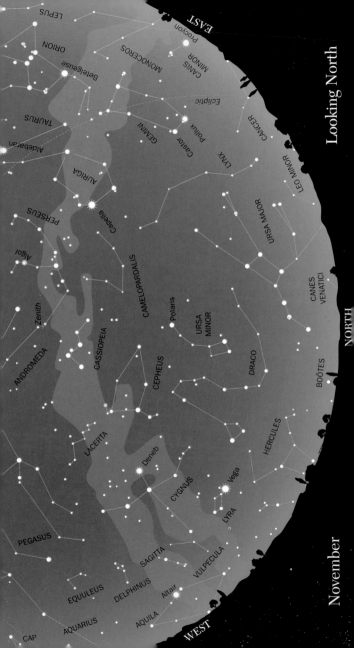

Gemini, with Castor and Pollux to the left

NOVEMBER
LOOKING NORTH

NOVEMBER CHARTS CAN BE USED AT THE FOLLOWING DATES AND TIMES:

Nov.01: 23:00 GMT
Nov.15: 22:00 GMT
Dec.01: 21:00 GMT

Winter is now truly upon us because Orion (p.41), the dominant constellation at this season, has reappeared over the eastern horizon. The constellation of **Gemini** (the Twins) is also now clearly visible, lying roughly parallel to the horizon in the east. Of the three stars that make up the Summer Triangle, Altair is very low in the west, but Deneb and Vega are still easily seen farther round towards the north. The small constellation of **Lacerta** (the Lizard) is well-placed for observation in the west, at an altitude that is slightly higher than Polaris. Auriga is high in the east, and both Andromeda and Perseus are right overhead on either side of the zenith.

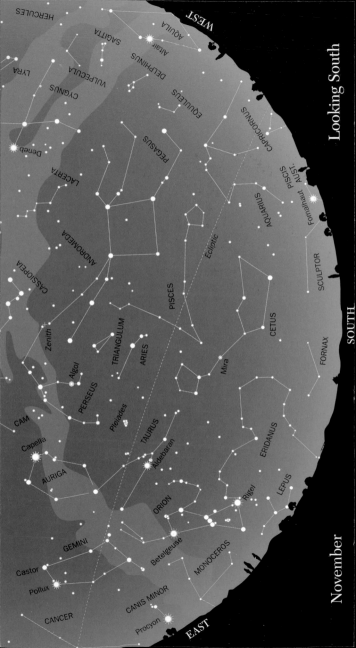

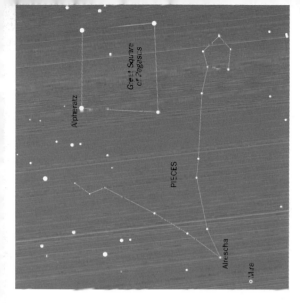

NOVEMBER
LOOKING SOUTH

South of the Great Square of Pegasus (now in the south-west) is a circlet of five stars. This is said to represent the body of the western fish in the zodiacal constellation of **Pisces** (the Fishes). The second (eastern) fish is not as distinct, consisting of a few faint stars to the east of the Great Square. The fish are shown (in mythological drawings) as being tied by their tails to two ribbons, marked on the sky by two long lines of stars, which form a 'V' to the south and east of Pegasus. The star **Alrescha** (α) at the apex of the 'V' represents the knot.

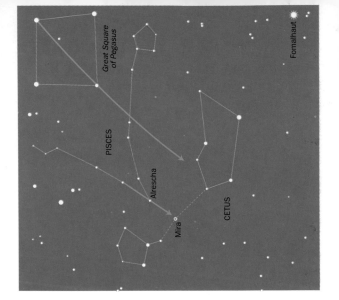

NOVEMBER

METEORS

Oct.20–Nov.30, maximum Nov.03:
Taurids – Slow meteors, frequently bright,
maximum hourly rate about 10.

Nov.15–20, maximum Nov.17:
Leonids – Fast meteors with trains, always interesting, but possibility of major displays in 1999 (see p.160).
Maximum rate possibly very high during that year

South of Pisces lies the constellation of **Cetus** (the Whale). This is not particularly easy to recognize. The 'V' of Pisces points straight to **Mira**, one of the most famous variable stars in the sky. When it is bright it is easily visible to the naked eye. When faint, the constellation may be treated as two parts. A diagonal through the Square of Pegasus point south-east towards the irregular polygon, with three fairly bright stars, that forms the 'body' of Cetus. The two bright stars in the pentagon that is the 'tail' lie to the east of Alrisha in Pisces.

Pisces, with Pegasus (top right) and Aries (left).

Pegasus, with the Great Square to the left

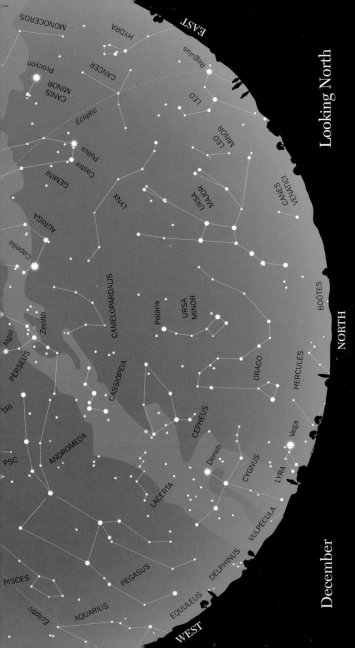

Perseus (p.230), with the Double Cluster (centre right)

DECEMBER
LOOKING NORTH

Sirius (p.48) has risen well above the horizon, so now all the brilliant constellations of winter are visible at once: Auriga, Taurus, Orion, Gemini, and Canis Major. The Summer Triangle, however, has not completely disappeared. Vega is skirting the horizon in the north-west, but Deneb, higher and farther west, is still clearly visible. Although nearly at the lowest part of its circuit, the head of Draco (and the rest of the constellation) is nicely placed for observation beneath Polaris. See if you can identify the faint constellation of **Lynx** (the Lynx) in the north-east. It runs generally north-south roughly half-way between the outer stars of Ursa Major and Castor and Pollux in Gemini.

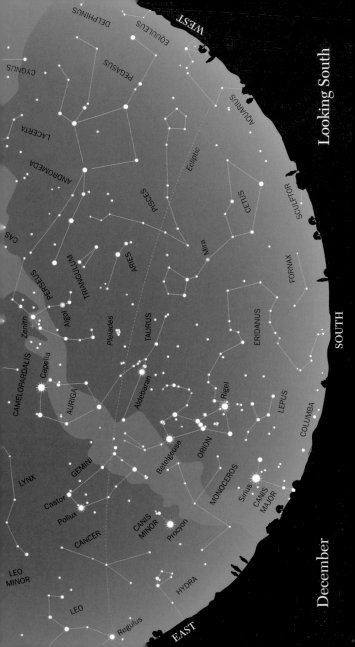

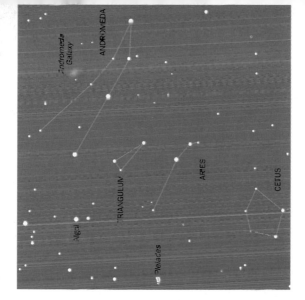

DECEMBER
LOOKING SOUTH

Beneath Andromeda to the west of the meridian are two small constellations. The first, **Triangulum** (the Triangle) is simply that: three not particularly outstanding stars, arranged in a tiny triangle. The other constellation, **Aries** (the Ram) is hardly any more impressive, despite being an important zodiacal constellation. The three brightest stars form a small, but distinctive, crooked pattern. Once learned, this is unlikely to be mistaken for any other similar asterism.

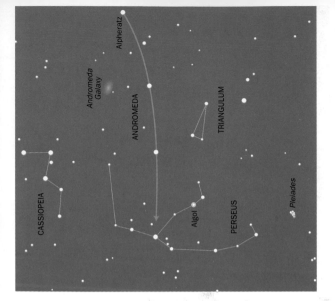

DECEMBER

METEORS

DECEMBER 7–16, maximum Dec.14 - **GEMINIDS**: a good shower with many bright, medium-speed meteors. Rewarding to photograph. **MAXIMUM HOURLY RATE** about 100

At the zenith lies the constellation of **Perseus** – which is probably easier to identify a month or so earlier, when it is lower in the sky. The bright chain of stars in Andromeda that runs east from the Square of Pegasus points to α Per, the brightest star, from which three lines of stars run up towards Cassiopeia, down towards the Pleiades (p.42), and back towards Triangulum. The last of these contains β Per, **Algol**, another famous variable star, which is normally bright. Even when it fades, it still remains easily visible to the naked eye, unlike Mira (p.102).

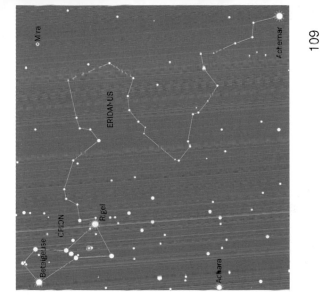

December

Finally, there is **Eridanus** (the river Eridanus) which starts close to Rigel in south-western Orion, and then winds its way to the west and south, disappearing beneath the horizon. It is an extremely long constellation and ends far to the south at the star Achernar, which is quite invisible from our latitudes.

THE PHASES OF THE MOON

Everyone is familiar with the changing appearance of the Moon throughout the month: **waxing** from New Moon to Full Moon, and then **waning** from Full Moon to New Moon again. These different **phases** arise from changes in the relative positions of the Sun, Moon and Earth, which mean that we see differing amounts of the hemisphere that is illuminated by the Sun. At New Moon, the Moon lies between the Earth and the Sun, with the dark side turned towards us, so it is invisible.

Following New Moon, a thin **crescent** is visible in the western sky for a short while after sunset. As the days pass, this crescent grows until (at First Quarter) half of the disk is illuminated. During the early crescent phases it is often possible to see a faint illumination on the 'dark' side of the Moon. This is **earthshine**, light from the Sun that has been reflected by the Earth onto the otherwise unlit side of the Moon. It may be quite bright when the relevant portion of the Earth is covered in clouds.

As the Moon moves eastward relative to the background stars, it remains visible later and later in the night. Between First Quarter and Full Moon, the Moon is **gibbous** (the visible disk is more than half illuminated). At Full Moon, the Moon is on the opposite side of the sky to the Sun, and is highest at midnight. Between Full Moon and Third Quarter the Moon is again gibbous, crossing the meridian after midnight. Eventually only a thin crescent is visible shortly before sunrise. We are back to New Moon again.

Because the Moon rotates once on its axis in exactly the same time as it takes to revolve once around the Earth, the same side of the Moon is always turned towards us. The Moon's orbit is not a perfect circle, however, but an ellipse, so the Moon is sometimes farther away from the Earth than at others, and its apparent diameter alters slightly. More importantly, its speed in orbit is not constant, so sometimes it is ahead of, and sometimes behind, the position it would have if its velocity were always the same. This causes the Moon to appear to rock backwards and forwards in the sky, so we occasionally see slightly farther round the eastern or western

Venus reappearing from occultation by the Moon

edge (or **limb**) of the Moon. As we saw earlier (p.25) the Moon is sometimes above the ecliptic, and sometimes below it, when we see more of the surface at the southern or northern limb, respectively. These movements (or **librations**) mean that, overall, about 59.5 of the Moon's surface is visible from Earth, although regions around the limb naturally appear greatly foreshortened.

Occultations

The Moon may pass in front of bright stars or planets in an event known as an occultation. the Moon is airless, so stars do not fade, but disappear and reappear instantaneously (unless they happen to be double stars (p.166), when their brightness changes in distinct steps). Because of the Moon's motion in declination (p.19), many years elapse between successive occultations of bright stars such as Regulus in Leo.

THE SURFACE OF THE MOON

The division of the surface of the Moon into light and dark areas is obvious even to the naked eye. In general, the light regions are the lunar **highlands**, which we now know consist of the oldest rocks on the Moon. The dark regions are known as **maria** ('seas'), and are low-lying plains. Some are actually giant impact basins that have been filled with dark lava flows. Nearly all are much younger than the highland regions, although they remain older than the majority of the rocks found on Earth.

Even binoculars reveal large numbers of the **craters** for which the Moon is famous, and the positions of some of the most conspicuous are shown on the following pages. Some craters (such as Plato) are conspicuous because they have been filled with dark lava, and there is a gradual transition to the smaller maria, such as Mare Crisium (which is actually circular, but appears elliptical because of foreshortening). Sinus Iridum on the edge of Mare Imbrium is a breached crater and many similar examples are visible.

Craters are easiest to see when their walls cast long shadows, around sunrise and sunset. For this they must be close to the line that divides the light and dark portions of the Moon. This line, the **terminator**, sweeps across the Moon's surface during the course of a month. For most craters there are favourable conditions for observation twice a month, but for the regions close to the lunar poles, and at the eastern and western limbs, it may be many months (or even years) before features are suitably illuminated at a time of maximum libration and thus easy visibility.

Because of the vertical lighting, not many features are clearly seen at Full Moon, although the highland/mare division is quite marked. The systems of **rays** that surround certain (relatively young) craters, however, are normally conspicuous. These streaks of fine ejected material are particularly bright around the craters Copernicus, Kepler, Aristarchus, and Tycho. The rays from Tycho (in the southern highlands) run right across the visible face of the Moon, and may be seen crossing Mare Serenitatis in the northern hemisphere.

The difference between the highlands and the maria is clearly visible

THE MOON: 3 DAYS OLD

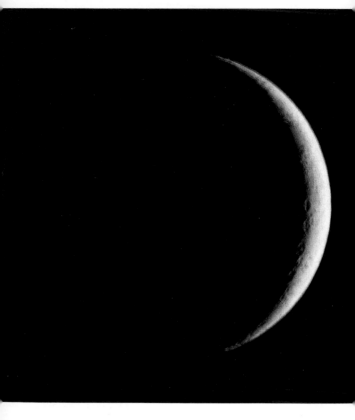

The Moon takes approximately 29.53 days to go through its complete cycle of phases (known as a **lunation**) from New Moon to New Moon. When a thin crescent – either waxing or waning – very few features are visible. Although some observers set out to see how soon after New Moon they can detect the hair-thin crescent, and have reduced the time to just a few hours, it is normally about three days before much detail can be detected with small telescopes or binoculars.

THE MOON: 3 DAYS OLD

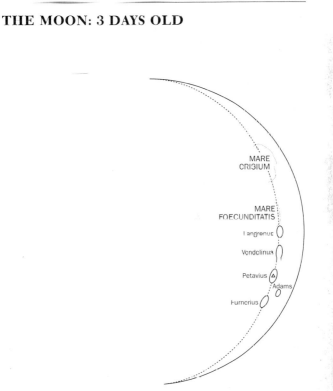

The three most conspicuous craters just on the terminator are (from top to bottom – North to South) Langrenus, Vendelinus, and Petavius, with its prominent central peak. Farther south, the rough wall of Adams may be seen, and the dark-floored crater Furnerius. North of Langrenus, the terminator crosses the flat areas on the edge of Mare Foecunditatis and Mare Crisium.

THE MOON: 7 DAYS OLD

Several maria are now prominent: Mare Crisium, Mare Foecunditatis, Mare Nectaris, most of Mare Tranquilitatis and part of Mare Serenitatis. Farther north, Lacus Somniorum and a portion of Mare Frigoris are also visible. The southern highlands are peppered with craters. Langrenus, Vendelinus and Petavius appear as bright spots under the high Sun. Near the terminator, the three adjacent craters, Theophilus, Cyrillus, and Catharina are extremely prominent. Nearby, Mare Nectaris extends south into the flooded crater Fracastorius. Farther south lies Piccolomini, with Lindenau and Zagut.

THE MOON: 7 DAYS OLD

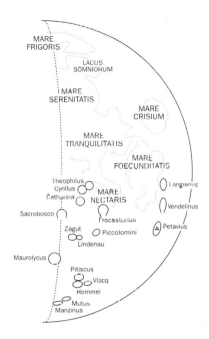

Farther south again lies the prominent group of Pitiscus, Vlacq, and Hommel. Mutus and Manzinus are in the south polar region. The terminator itself runs through the craters Sacrobosco and Maurolycus.

THE MOON: 10 DAYS OLD

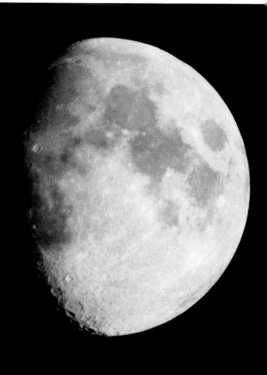

The whole of Mare Serenitatis is now visible, with Plinius, Menaelaus, and the Haemus Montes on its southern border. A bright ray crosses the centre of the mare. Farther west, most of Mare Imbrium may be seen, with Montes Jura on the terminator. The dark-floored crater Plato is surrounded by the Montes Alpinus. Farther south the mare is bordered by the Montes Apenninus, with Eratosthenes at the end of the chain. Archimedes is visible within the mare itself. The brilliant crater Copernicus lies between Mare Imbrium and the larger Oceanus Procellarum. Farther south, there is Mare Nubium

THE MOON: 10 DAYS OLD

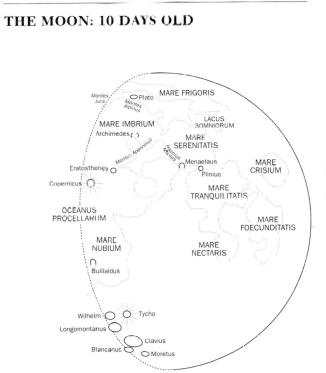

and the large crater Bullialdus. Most of the rays spreading over the disk originate in the crater Tycho. To its west, near the terminator, are Wilhelm and Longomontanus. Farther south lie Clavius, Blancanus and Moretus.

THE MOON: 14 DAYS OLD

At Full Moon, we are looking directly down onto the surface in the same direction as the Sun's rays, so nearly all the shadow detail is lost, except around the limb, where all the features are foreshortened, and difficult to recognize. At this period, however, the ray systems are particularly prominent, especially that around Tycho, which also displays its dark aureole. Other ray systems surround Copernicus and Kepler. In the western Oceanus Procellarum, the crater Aristarchus is a prominent bright feature. Close to the western limb is the dark-floored crater Grimaldi. In the centre of the disk are the

THE MOON: 14 DAYS OLD

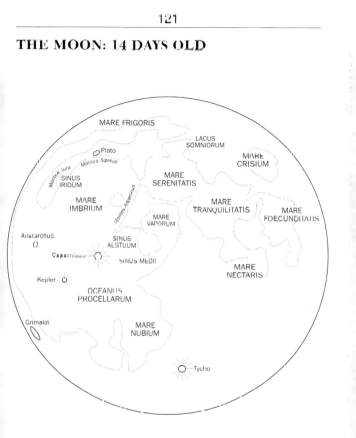

dark areas of Mare Vaporum, Sinus Aestuum, and Sinus Medii.

THE MOON: 18 DAYS OLD

The evening terminator has overtaken Mare Crisium and Mare Foecunditatis but features around Mare Serenitatis and Mare Tranquilitatis are now easier to see, particularly Posidonius. The crater Theophilus is prominent on the borders of Mare Tranquilitatis and Mare Nectaris. Farther south lies the large crater Maurolycus, with Tycho to its west. The triple craters of Pitiscus, Vlacq, and Hommel lie close to the terminator. Slightly to the north, right on the terminator, lie Janssen and Fabricius. In the west, Grimaldi and Aristarchus are still prominent. In the northern part of the

THE MOON: 18 DAYS OLD

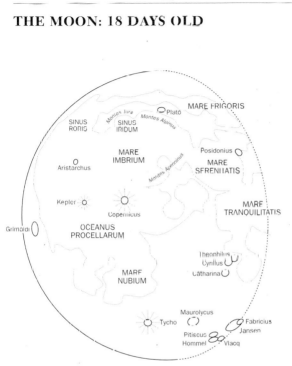

disk, practically the whole extent of Mare Frigoris is visible, together with Sinus Roris. The Mare Ibrium, which, together with its surrounding mountains, were all created by a giant impact, is readily seen at this stage of the lunation.

THE MOON: 22 DAYS OLD

The terminator has now moved so far across the disk that most of the cratered highlands are hidden, and the dark, low-lying Mare Imbrium, Oceanus Procellarum and Mare Humorum are dominant. Aristarchus, on the edge of Mare Imbrium is bright, as is the area around the crater Crüger in the cratered terrain separating Oceanus Procellarum and Mare Humorum. Sinus Iridum and Copernicus and, farther south, Grimaldi are thrown into relief by the advancing shadow. Bullialdus is still visible as, farther to the south, are Maginus and Clavius. Tycho is right on the terminator. The crater

THE MOON: 22 DAYS OLD

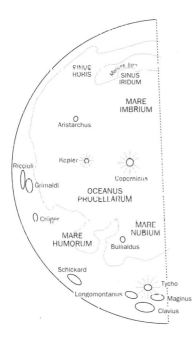

Grimaldi is easily seen and a dark spot near it is part of the floor of Riccioli. Schickard, too, farther south, also has a dark floor.

THE MOON: 25 DAYS OLD

At this phase the terminator is approaching the western limb of the Moon and very few features are visible. Mare Humorum is now in shadow and only the relatively featureless edges of Oceanus Procellarum and Sinus Roris remain visible. The dark floors of Grimaldi, Riccioli and Schickard may be seen, together with various brighter patches on the highland terrain. Depending on the degree of libration (p.111), other craters may become visible in the region along the limb, especially in the south. Almost due west of Aristarchus, the

THE MOON: 25 DAYS OLD

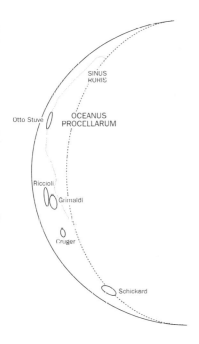

large crater Otto Struve is often seen at about this time in the lunation.

ECLIPSES

If the Moon's orbit around the Earth were exactly aligned with the ecliptic (the Earth's orbit around the Sun), there would be a **solar eclipse** at every New Moon, when the bodies were in line, and a **lunar eclipse** at every Full Moon. This does not occur, because the orbit is inclined, and variations occur in both the inclination and the orientation in space. (The inclination varies between approximately 18° and 28.5°.)

As a result, the number and type of eclipses in any one year vary considerably. The minimum number of eclipses is two (both of which must be solar), and the maximum seven. There is a long cycle (known as the **Saros**) of 18 years and 10.3 or 11.3 days after which a similar pattern of eclipses recurs. The repetition is not exact, however, and the paths of totality of the solar eclipses slowly drift southwards, as part of a longer cycle that lasts approximately 1,262 years.

Solar eclipses

Solar eclipses occur when the Moon comes between the Sun and the Earth (at New Moon) and its shadow is cast on the Earth's surface. The dark, central shadow cone is known as the **umbra**, and the much larger outer region where part of the Sun's disk is hidden is the **penumbra**. By coincidence, Moon and Sun appear almost exactly the same size (approx.0.5° in diameter), as seen from Earth, so the tip of the umbra

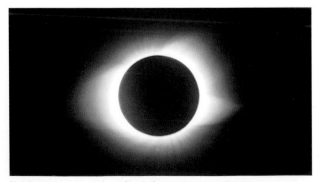

The Sun's corona varies markedly from eclipse to eclipse

sometimes reaches the Earth's surface and the Sun's disk is completely obscured by the Moon. We then have a **total eclipse**, the most spectacular type, when the Sun's outer atmosphere, the **corona**, with its plumes and rays, becomes visible.

As we have seen earlier (pp.24–25), the distances between the Earth and the Moon, and between the Earth and the Sun both vary. The most favourable conditions occur when the Moon is closest to the Earth (at **perigee**), and the Earth is farthest from the Sun (at **aphelion**). The Moon appears largest and the Sun smallest, so the duration of the eclipse is longest and theoretically it could last 7 minutes 31 seconds. Such ideal conditions occur rarely and most total eclipses are much shorter.

Eclipses may happen when the Moon is farthest from the Earth (at **apogee**) and the Earth is closest to the Sun (at **perihelion**). The Moon's disk no longer covers the Sun (i.e., the umbra does not touch the Earth), and a ring of light remains visible, giving an **annular eclipse**.

Under the most favourable conditions with the Sun and Moon directly overhead, the maximum diameter of the umbra is 273 km. Generally, the diameter is much less: 150–160 km. Towards the poles the diameter increases and may occasionally grow to 780 km. The combination of the Earth's rotation and the Moon's motion in its orbit causes the shadow to race eastwards across the Earth at about 3,200 km/h. Totality thus occurs within a narrow band running across the globe, which is why total eclipses are rare at any one place. Anyone outside this narrow path and within about 3,200 km of the central line will see a **partial eclipse**.

Warning: Never look at the Sun, at any time, with any binoculars or telescope. You risk serious eye damage. Even low on the horizon, when the Sun appears dim, infrared radiation (which you cannot feel) may still be strong enough to injure your retina. So-called 'solar filters' that fit inside telescopes should never be used. Many allow damaging radiation to pass; others may fracture from the concentrated heat. The only safe way to view the Sun is by projecting an image onto a card that is held behind the eyepiece, or by using a proper, specialized solar filter, which reflects most of the light and heat away **before** it enters the telescope. Even then you should not use a finder to align the telescope. (In fact it should be covered to prevent light from entering it.)

Lunar eclipses

Unlike solar eclipses, eclipses of the Moon may be seen from anywhere on the Earth's hemisphere that faces the Moon. They also last much longer, and the maximum duration of totality is 107 minutes. The Moon's eastward motion against the sky carries it into the penumbra, through the dark umbra and out into the penumbra again. Generally, the penumbral dimming is so slight that it is unnoticed, and it is only when the Moon reaches the umbra that you realize that an eclipse is taking place. Penumbral eclipses, when the Moon does not enter the umbra, are of little interest to most observers.

Even in the umbra, the Moon does not normally disappear, although its brightness and colour vary considerably at each eclipse. Conditions in the Earth's atmosphere cause the differences. During totality the Moon is normally illuminated by a coppery light that has been refracted into the shadow cone by our atmosphere. There may also be a bluish tinge to the edge of the umbra. Only rarely does the Moon become so dark that it disappears at mid-eclipse. Normally there is enough light to be able to make out the major surface features.

Table of solar eclipses: 1999–2003

1999	Feb.16	Annular	0^m40^s	S Africa, Antarctica, Australasia
	Aug.11	Total	2^m23^s	Arctic, Europe, N Africa, Arabia, Asia
2000	Feb.05	Partial		Antarctica
	Jly.01	Partial		SE Pacific, SW South America
	Jly.31	Partial		NE Asia, Alaska, Canada, Greenland
2001	Jun.21	Total	4^m57^s	South America, S & Central Africa
	Dec.14	Annular	3^m53^s	Hawaii, SW Canada, W USA, Mexico, Caribbean
2002	Jun.10 11	Annular	0^m23^s	SE Asia, Philippines, North America (not N and E)
	Dec.04	Total	2^m04^s	S Africa, Madagascar, S & W Australia
2003	May.31	Annular	3^m37^s	Arctic regions, Greenland, Iceland, N Europe, N Asia, Alaska, N Canada
	Nov.23–24	Total	1^m57^s	Antarctica, S Australasia, S South America

Table of lunar eclipses: 2000–2003 (none in 1999)

2000	Jan.21	Total	NW Asia, N & S America, Europe, N & W Africa
	Jly.16	Total	Pacific, Antarctica, Australasia, SW Asia
2001	Jan.09	Total	Australia, Indonesia, Philippines, Asia, Africa
2003	May.16	Total	Antarctica, Africa, Europe, S & most of N America
	Nov.08-9	Total	W Asia, Europe, N America

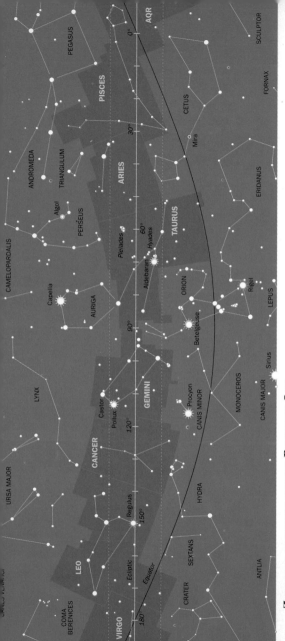

ZODIACAL CONSTELLATIONS: PISCES TO LEO

The zodiacal constellations are shown here by the darker tint. The Sun always lies on the ecliptic, and the Moon and planets lie within the region indicated by the dotted lines. Note how this zone also contains parts of several other constellations, such as Auriga, Cetus, Opiuchus, Orion, and Sextans.

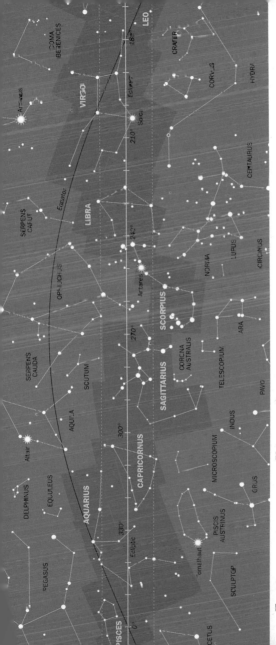

ZODIACAL CONSTELLATIONS: VIRGO TO AQUARIUS

This projection shows the ecliptic as a straight line, and the celestial equator as curved. Ecliptic longitude is reckoned eastward along the ecliptic, beginning at 0°, the vernal equinox, or First Point of Aries (p.24), now in Pisces.

133

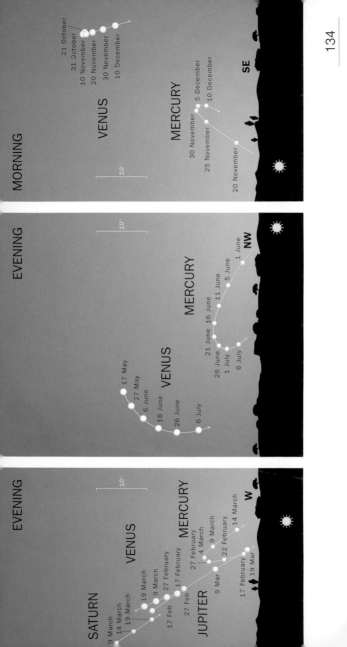

EVENING

SATURN
9 March
14 March
19 March

VENUS
27 February
9 March
19 March

MERCURY
27 Feb
17 Feb
17 February
27 February
4 March
9 March
14 March
22 February
9 Mar

JUPITER
17 February
19 Mar

W

10°

EVENING

VENUS
17 May
27 May
6 June
16 June
26 June
6 July

MERCURY
21 June
16 June
11 June
26 June
5 June
1 July
6 July
1 June

NW

10°

MORNING

VENUS
21 October
31 November
10 November
20 November
30 November
10 December

MERCURY
30 November
25 November
20 November
5 December
10 December

SE

10°

134

PLANETARY POSITIONS 1999

Mercury is easiest to see in the evening sky during late February and early March – when three other planets (Venus, Jupiter and Saturn) are in the same region – and in the morning in late November and early December. It is also visible in the evenings in June, but is then closer to the horizon. As seen from Earth, it actually crosses (or transits) the disk of the Sun on November 15. Note that appropriate safety precautions (see p.129) **must be** taken before attempting to observe such an event.

Venus is a prominent object in the evening sky from February until July, when it begins to approach too close to the Sun to be visible. It reappears in the morning sky later in the year and is extremely bright (mag. –4.4) when at greatest elongation on October 31.

Mars begins the year in Virgo, and is in the same constellation at its opposition on April 24, when it appears brighter (mag. –1.7) than Sirius. It moves eastwards through Libra, Scorpius, and Ophiuchus during the summer and early autumn twilight, finally becoming visible in dark skies again at the end of the year.

Jupiter disappears into evening twilight at the end of February, reappearing in the morning sky in June. It is at opposition in Pisces at mag. –2.9 on October 23. As the year passes it slowly overtakes Saturn in the same region of sky.

Saturn begins the year in Pisces and remains visible until March. Like Jupiter it reappears in June/July in Aries, where it is at opposition on November 6 at mag. –0.2.

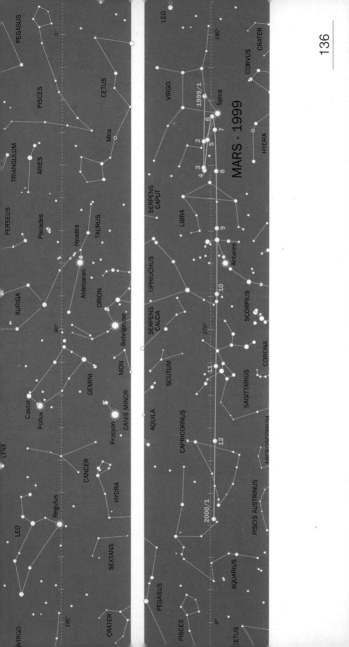

MARS - 1999

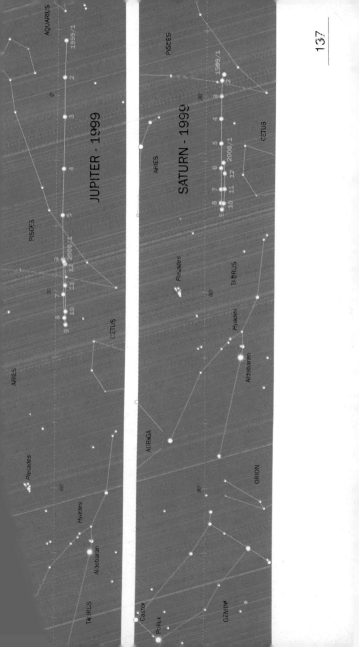

JUPITER - 1999

SATURN - 1999

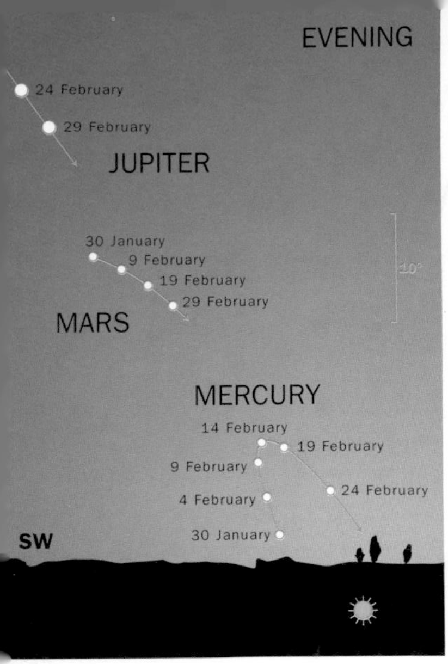

EVENING

JUPITER

24 February
29 February

30 January
9 February
19 February
29 February

MARS

MERCURY

14 February
19 February
24 February
9 February
4 February
30 January

SW

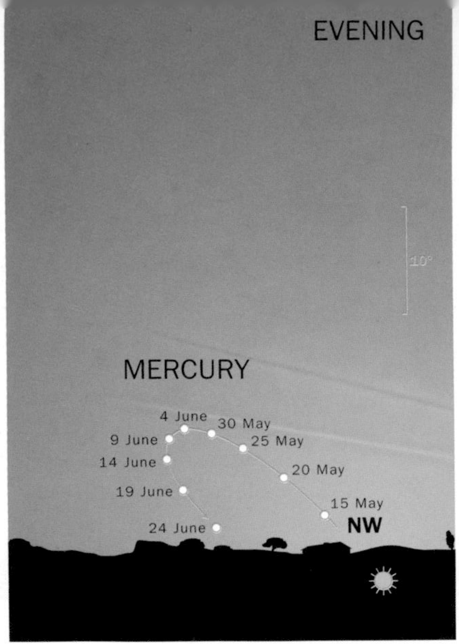

EVENING

MERCURY

4 June
30 May
9 June
25 May
14 June
20 May
19 June
15 May
24 June

NW

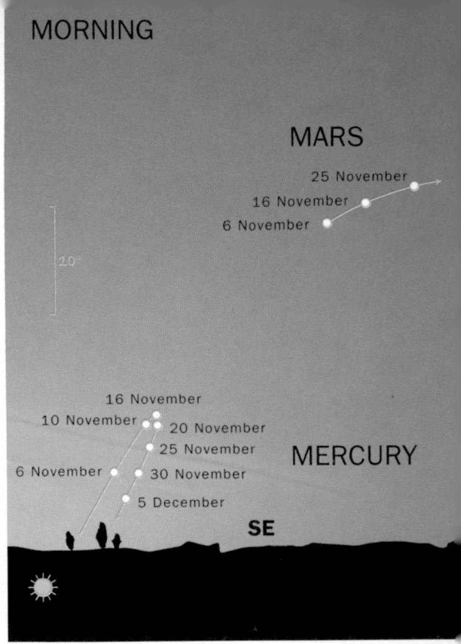

MORNING

MARS

25 November
16 November
6 November

MERCURY

16 November
20 November
10 November
25 November
6 November
30 November
5 December

SE

10°

138

PLANETARY POSITIONS 2000

Mercury is in the evening sky in the middle of February and mid-June, but easiest to see in the early morning in mid-November. Although brighter in January and May, it is then very close to the horizon.

Venus's not very easily visible this year, although it begins to appear in the evening twilight in late November and December.

Mars does not come to opposition in 1999. It may be seen early in the year, in Aquarius and Pisces, until it disappears in twilight in April. It reappears in the morning sky in September in Leo, and becomes increasingly prominent towards the end of the year.

Jupiter, like Mars, is reasonably placed in the evening sky for the first couple of months, but it dominates the night sky later in the year, especially around opposition on November 28 (in Taurus) when it reaches mag. -2.9.

Saturn slowly moves from Aries into Taurus during the course of the year, and has a similar period of visibility to Jupiter, being best seen in the latter half of the year. It is at opposition on November 20, at mag. -0.9.

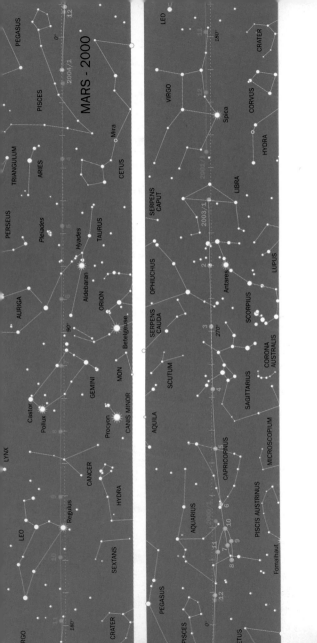

MARS - 2000

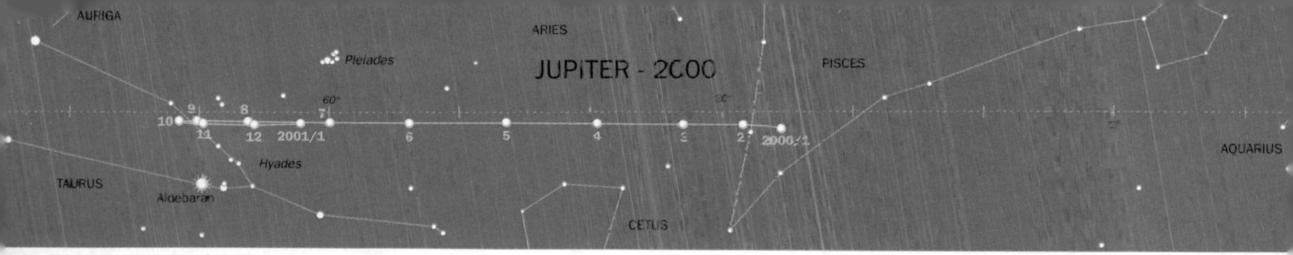
JUPITER - 2000
AURIGA
ARIES
Pleiades
PISCES
AQUARIUS
TAURUS
Aldebaran
Hyades
CETUS
60°
30°
2000/1
2001/1

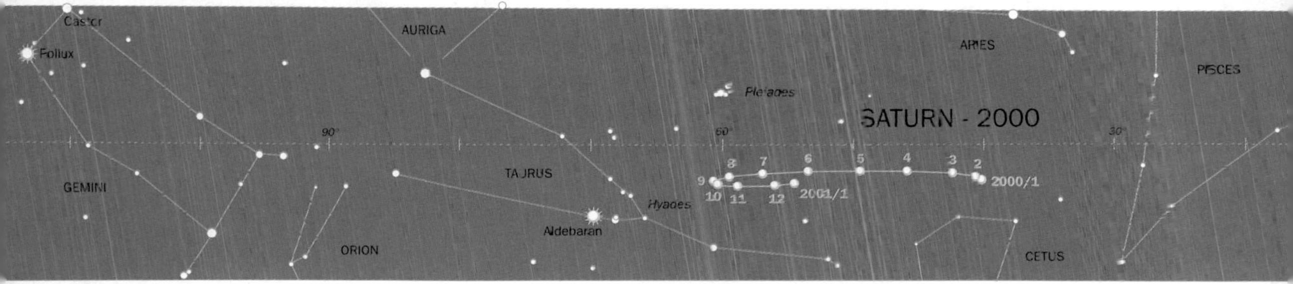
SATURN - 2000
Castor
Pollux
AURIGA
ARIES
PISCES
Pleiades
GEMINI
TAURUS
ORION
Aldebaran
Hyades
CETUS
90°
60°
30°
2000/1
2001/1

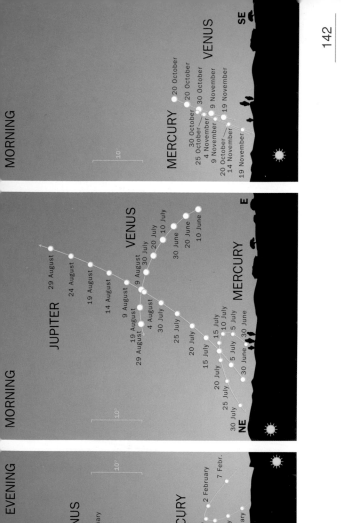

EVENING

VENUS
2 February, 12 February
22 February
23 January
13 January
3 January

MERCURY
28 January
23 January
2 February
18 January
7 Febr.
13 January
SW

MORNING

JUPITER
29 August
24 August
19 August
14 August
9 August
9 August
19 August
4 August
29 August
30 July
25 July
20 July
15 July
20 July
25 July
30 July
NE

VENUS
30 July
20 July
30 June
20 June
10 July
10 June

MERCURY
15 July
15 July
10 July
5 July
5 July
30 June
30 June

MORNING

MERCURY
20 October
20 October
30 October
30 October
25 October
4 November
4 November
9 November
9 November
9 November
19 November
20 October
14 November
19 November
VENUS
SE

PLANETARY POSITIONS 2001

Mercury is easiest to see in late January and late October, in the evening and morning skies, respectively. It may be glimpsed with difficulty in the morning in mid-July. On all three occasions, Venus is also in the same region of the sky.

Venus is prominent in the evening sky in January and February (when it reaches mag. -4.7), and again in October, although it will then be closer to the horizon. It is reasonably easy to see in the middle of the year, when its magnitude will be around 4.0.

Mars is visible at some time of the night throughout most of the year. At opposition on June 14, it is on the borders of Ophiuchus and Sagittarius, almost at the lowest part of the ecliptic, but will then be in twilight. It is most readily seen early in the night late in the year.

Jupiter does not come to opposition in 2001, but it is visible for a large part of the night until April. It reappears in the morning sky in late July, and by December is visible throughout the night.

Saturn is also visible for a large part of the year. It is moderately bright in the early months (about -0.1 mag.) but reaches -0.4 at opposition on December 4, when it is in Taurus.

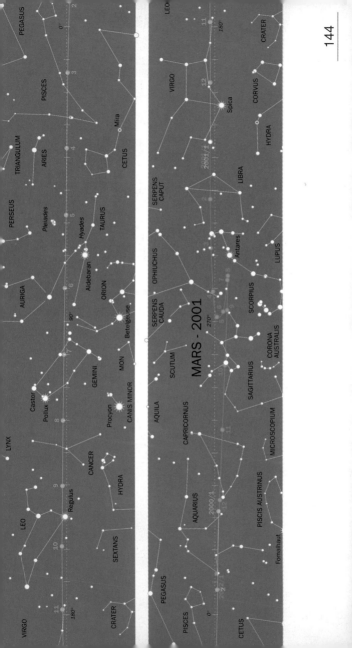

MARS - 2001

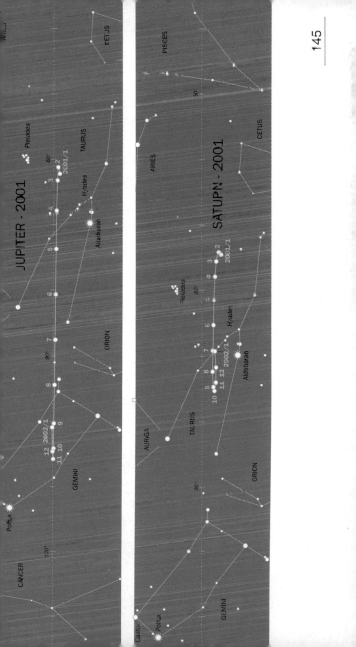

JUPITER - 2001

SATURN - 2001

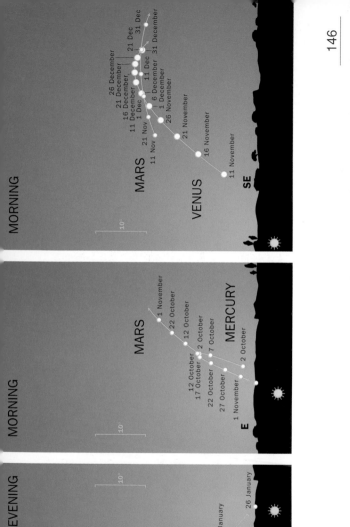

EVENING   MORNING   MORNING

MERCURY

11 January
5 January
1 January
16 January
21 January
26 January

SW

MARS

1 November
22 October
12 October
12 October
17 October
2 October
7 October
22 October
27 October
1 November
2 October

MERCURY

E

MARS

26 December
21 December
16 December
11 December
6 December
1 Dec
21 Dec
31 Dec
11 Dec
6 December
1 December
11 Nov
21 Nov
11 Nov
26 November
21 November
16 November
11 November
31 December

VENUS

SE

10°

PLANETARY POSITIONS 2002

In a rare conjunction, all five planets Mercury, Venus, Mars, Jupiter and Saturn are visible in the western sky, accompanied by the crescent Moon, between May 13 and 16.

Mercury is visible in the west in January, but the morning apparition in mid-October is likely to be more favourable.

Venus is a brilliant morning object from early November until the end of the year. It reaches a magnitude of about -4.7 in early December.

Mars is an evening object in the first quarter of the year. It is invisible from the end of April until October. It does not have an opposition in 2000, so never becomes very bright

Jupiter is at opposition on January 1 at about mag. -2.6, but declines to about -2.0 before disappearing into twilight in late May. It reappears after midnight in August, brightening and becoming visible longer each night until the end of the year.

Saturn is in Taurus, to the west of Jupiter, and both disappears (in early May) and reappears (in July) slightly earlier than the larger planet. From a minimum of about mag. +0.1 it brightens to about -0.5 at opposition on December 18.

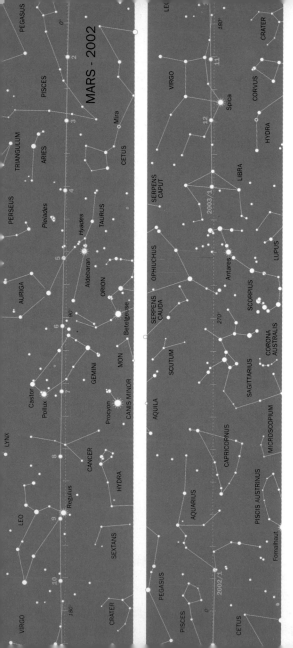

MARS - 2002

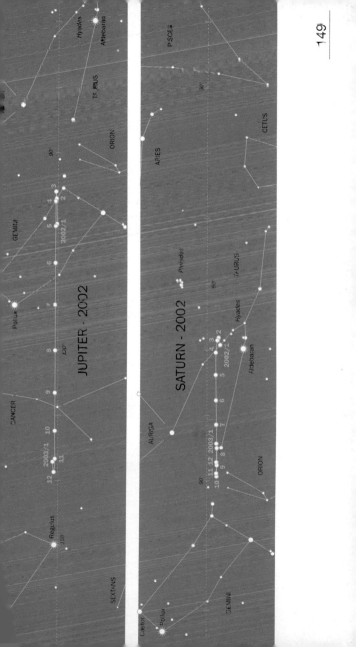

JUPITER - 2002

SATURN - 2002

149

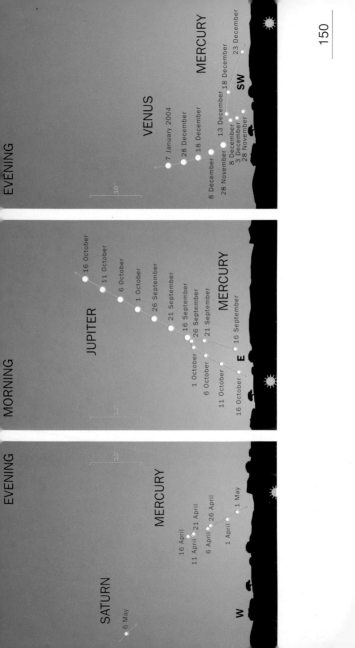

EVENING

SATURN

● 6 May

MERCURY

16 April
11 April ● 21 April
6 April ● ● 26 April
1 April ● ● 1 May

W

MORNING

JUPITER

● 16 October
● 11 October
● 6 October
● 1 October
● 26 September
● 21 September
● 16 September
● 26 September
● 21 September
1 October ● ● 16 September
6 October ●
11 October ●
16 October ●
MERCURY
E

EVENING

VENUS

● 7 January 2004
● 28 December
● 18 December

8 December ●
28 November ● ● 13 December
● 8 December
● 3 December
28 November ● MERCURY
SW
● 18 December
● 23 December

PLANETARY POSITIONS 2003

Mercury has three apparitions, those for mid-April and September/October being the most favourable, while that for December is rather low on the horizon.

Venus is best seen either as a morning object early in the year, or on December evenings, when it is brightening rapidly, and getting farther above the horizon. (Visibility is exceptionally good in 2004.)

Mars may be seen for several months early in the year, but tends to be close to the horizon. Visibility rapidly improves from early July, and opposition is on August 29, at mag. -2.9.

Jupiter has an extended period of visibility early in the year, with opposition in Cancer on February 2 at mag. -2.6. It reappears in the morning sky in September, and rises around midnight at the end of the year.

Saturn is visible for most of the night early in the year, but slowly fades, and is lost in twilight by May. It reappears in the early morning in July and by early December is visible throughout the night once more. There is no opposition in 2003.

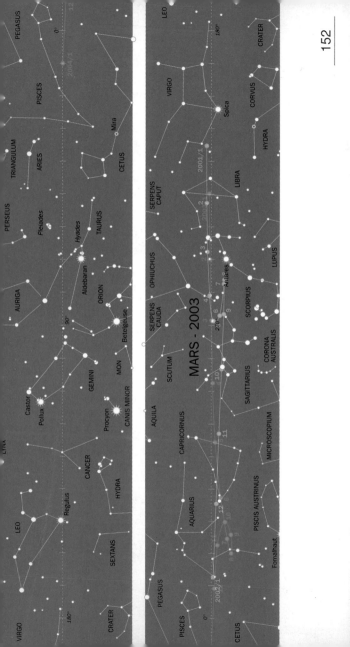

MARS - 2003

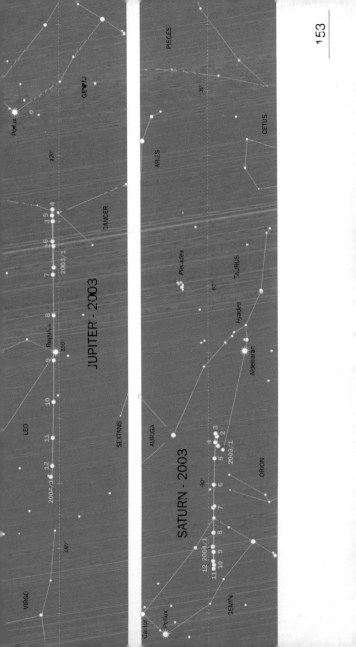

JUPITER - 2003

SATURN - 2003

AURORAE

A major auroral display is an unforgettable sight. Unfortunately they are infrequent at our latitudes, but become more common the farther north you go. In Europe, Norway, Sweden and Finland are particularly favoured – not least because of the long winter nights – but displays are also regularly seen from Scotland. The frequency of aurorae varies in accordance with the sunspot cycle of approximately 11 years, but a major auroral storm, such as the one that occurred on 13 March 1989, and which was visible from most of the northern hemisphere, takes place perhaps two or three times a century.

An aurora occurs when the Earth's magnetic field captures energetic particles emitted by the Sun, accelerates them to high velocities, and causes them to cascade into the upper atmosphere in regions called the auroral ovals, roughly centred on the Earth's magnetic poles. Here, generally at heights of 100–300 km, the energetic particles collide with atoms (particularly oxygen and nitrogen), and cause them to emit light, which produces the visible aurora.

Most auroral displays start either as an indistinct **veil** of light, or more concentrated **patches**. A common form is an **arc**, an arch of light centred in the north, with a sharp lower border and a more diffuse upper one. Frequently, vertical **rays** then develop, which may extend high into the sky, either as part of a rayed arc, or as isolated rays. During a strong display, the arc may widen and become a distinct, broad **band**, which may also show a rayed structure. Sometimes a band will develop folds, which may move around like an enormous curtain waving in the wind. Multiple bands and curtains are seen in major displays. Occasionally a display may appear overhead, when the rays appear to converge at a point high in the sky. This is known as a **corona**. During an active display, the appearance may change dramatically from one moment to the next, with surges and flares of activity, which often move rapidly across the sky.

The colours visible in an aurora depend greatly upon individual observers' eyesight. The most commonly reported colour is green and this is the normal colour of the lower edge of an aurora. Red often appears higher up, sometimes as striking red rays, but some people have difficulty in seeing this particular wavelength. During an intense display, some observers may see purple-violet tints where the light is particularly strong.

Aurorae are quite easy to photograph with a camera mounted on a tripod. With a film having a speed of 400 ISO, and an aperture of f/1.8, exposures of 15–30 seconds should give good results. If the display is changing rapidly, shorter exposures may be needed to avoid blur in the final photograph. It is a good idea to include part of the foreground in the picture. This can help to determine the direction and angular size of parts of a display. Most auroral photographs show background stars, which are also useful in this respect. As with all photographs of the night sky, remember to record the time as accurately as possible.

Green and red are the most common auroral colours

NOCTILUCENT CLOUDS

The short nights around the summer solstice, when twilight persists throughout the night, are not ideal for observing faint stars and deep-sky objects. The farther north you are, the worse the situation becomes with a strong twilight arch in the north even at midnight. As some compensation, however, this is just the time of year (from late May to mid-July) when you may see noctilucent clouds (NLC) shining in the northern skies. These silvery-white or bluish-white clouds appear somewhat like ordinary cirrus clouds, and displays are usually seen several times during a season.

They are the highest clouds in the Earth's atmosphere and lie at the top of the region known as the mesosphere, at about 82 km. This is several times as high as ordinary clouds, which may frequently be seen as dark silhouettes against the twilight arch and the much higher NLC.

Noctilucent clouds are visible only when you are in darkness, but the clouds themselves are still in daylight, illuminated by the Sun from below your northern horizon. For this reason there is a southern limit to their visibility, but anyone north of about 50°N latitude may be able to see them at some time during the summer. As you go north your chances of seeing a display increase. The clouds occur in a very thin layer, however, so they are easiest to see when your line of sight grazes the cloud sheet. When directly overhead they may be too thin to be readily visible.

The clouds usually show some form of structure, such as waves and ripples, although they sometimes appear as a featureless veil. Their appearance changes with time, partly because of the changing direction of illumination during the night, but also because the clouds move in response to the high-level winds. This region of the atmosphere is poorly understood, so observations by keen amateurs are of considerable scientific value. The nature of the clouds themselves was in some doubt for many years, but we now know that they consist of water ice, although the precise conditions that control their formation are still uncertain.

NLC are easy to photograph with ordinary cameras and lenses. Using colour film with a speed of 200 ISO, exposures of about 20 seconds are suitable around midnight. Earlier and later in the night, the exposure should be shorter, down to about 2 seconds at the end of civil twilight in the evening and the same at the beginning of civil twilight at dawn. If you would like your photographs to be of scientific use, keep the camera pointing in a fixed direction on its tripod. The motion of the clouds may then be seen from one frame to the next. If the same display is photographed by another observer, the cloud heights and wind speeds may then be determined from simple trigonometry and a knowledge of the observers' positions. If possible, make exposures on the hour, and at 15, 30 and 45 minutes past. Again, this makes it possible to compare your photographs with those taken by other people at exactly the same time.

Highly characteristic structure in noctilucent clouds

METEORS

Nearly everyone who has ever looked up at the stars has seen a **meteor** (or 'shooting star'), flit across the night sky. Although some meteors may be quite bright, most are actually caused by tiny particles of interplanetary dust. Very few are larger than a pea, but they encounter the tenuous gases of the Earth's upper atmosphere at such great speeds (11–72 km/s) that their outer layers are rapidly vaporized and give rise to the luminous meteor. Most meteors occur at heights of about 70–100 km.

The names of these objects sometimes cause confusion. When out in space, the particles are known as **meteoroids**, regardless of their size, and become meteors when they begin to glow. If a portion survives to reach the surface, it is known as a **meteorite**. The smallest meteoroids (less than about 0.1 mm in diameter and with masses of about one millionth of a gram), are not vaporized. These **micrometeorites** slowly drift down to the surface and may be captured by high-altitude aircraft or balloons, or else recovered from cores drilled into the polar ice-caps or deep-sea sediments.

Much larger bodies (with masses greater than about 1 kg) give rise to meteorites, and these may weigh anything from a few grammes to many tonnes. In most cases, the body loses its initial velocity as it is braked by the atmosphere, and finally

A special shutter produced the breaks in the trail of this bright fireball

falls almost vertically, creating a small impact crater where it lands. Such meteorites (of which there are many types) are of great scientific importance, because they provide information about the early stages of the formation of the Solar System, or even about conditions that existed before it was formed (4,650 million years ago). A number of meteorites are known to have come from the Moon and from Mars, but most are thought to be fragments of the small bodies known as minor planets that orbit the Sun alongside the larger planets. Most visible meteors (as opposed to meteorites) are caused by particles shed by comets as they travel through the Solar System.

Much larger bodies (1,000 tonnes or more) are practically unaffected by the atmosphere, and hit the surface with most of their initial velocity, creating an explosion crater, like the famous Meteor Crater in Arizona, or the craters on the Moon.

The brightness of meteors is given in magnitudes, like that of stars (p.29). It takes experience to judge the magnitude accurately, but observations by advanced amateurs are used to determine the approximate sizes and numbers of meteors encountering the Earth's atmosphere at particular times of the year. Some meteors leave persistent **trains** behind them, marking their paths, and observation of these can provide information about upper-atmosphere winds.

An extremely bright meteor of mag. -5 or more, brighter than the brightest planets (Venus or Jupiter), is known as a **fireball**. Some may be bright enough to cast shadows, and even exceed the brightness of the Full Moon (mag. -13). These fireballs are important, because some may give rise to meteorites, which should be recovered as soon as possible after they fall to be of greatest scientific value.

Most meteors occur so high in the atmosphere that the sounds that they make (if any) are undetectable. Occasionally, however sounds may accompany bright fireballs (which are then known as **bolides**), particularly those that end as meteorites. There is usually a sonic boom, caused because the body is travelling faster than the speed of sound, and there may be explosions if it breaks into smaller fragments. If you divide the time in seconds between seeing the bolide and hearing any sounds by three, this gives you an estimate of the distance in kilometres.

Meteor showers

Many meteors appear from random directions in the sky, and these **sporadic meteors** occur throughout the year. More interesting, however, are the meteors that have been shed by comets. These become spread out along the comets' orbits and encounter the Earth at specific times of the year, to give rise to a **meteor shower**. Such showers may last for several days and usually begin at a low rate, rise to a peak, and then decline in numbers. The table gives details of some important showers, and you will see more meteors if you observe when these showers are active. Some meteors are not spread evenly around their parent comet's orbit, but are bunched together, giving rise to major showers or meteor storms at regular intervals. The most famous are the Leonids which have a 33-year period, like their parent comet Tempel-Tuttle. A rate of 140,000 per hour was estimated for the intense short peak in 1966. Rates were about 250 per hour in 1998, and are expected to be much higher in 1999.

A shower's meteoroids all travel in parallel paths in space, but when they encounter the Earth, because of perspective, their trails appear to diverge from a single area of the sky. This is known as the **radiant**, and particular meteor showers are named after the constellation in which the radiant lies. The Leonids, which diverge from Leo, and the Perseids (from Perseus) are just two examples. One shower, the Quadrantids, is named after an old constellation, Quadrans Muralis (the Mural Quadrant), no longer used. The radiant lies at the top of Boötes, near the border with Draco.

How do you decide whether a meteor belongs to a particular shower or is a sporadic? The simplest method is to use a piece of string, or a long, straight stick. Hold the string or stick up along the path where you saw the meteor. If the path, extended backwards, passes within 4° of the radiant position, then you may safely assume that the meteor belonged to that particular shower. The positions of the major shower radiants are marked on the individual constellation charts.

During the time when a shower is active, you will see most meteors if you do not look directly at the radiant, but instead watch the sky about 45° away (and about 45° above the horizon).

Approximate dates of meteor showers

NAME	MAXIMUM	LIMITS	RA	DEC	HOURLY RATE
Quadrantids	Jan.04	Jan.01–06	15 28	+50	100
Lyrids	Apr.22	Apr.19–25	18 08	+32	10
η Aquarids	May.05	Apr.24–May.20	22 20	-01	35
δ Aquarids	Jly.28	Jly.15–Aug.20	22 36	-17	20
(double radiant)			22 04	+02	10
Perseids	Aug.12	Jly.23–Aug.20	03 04	+58	80
Orionids	Oct.21	Oct.16–26	06 24	+15	25
Taurids	Nov.03	Oct.20–Nov.30	03 44	+14	10
Leonids	Nov.17	Nov.15–20	10 08	+22	??
Geminids	Dec.13	Dec.07–15	07 28	+32	100

ARTIFICIAL SATELLITES

Artificial satellites are visible when the observer is in darkness, and the satellites are illuminated by the Sun. Such conditions occur shortly after sunset or before dawn (or even throughout the night during summer). Satellites may flash as they tumble and catch the sunlight, but at a much slower rate than the rapid flashing from the lights of high-flying aircraft.

Predictions for the brightest satellites and for the Space Shuttle or MIR manned craft are published in many newspapers, although most Space Shuttle missions cannot be seen from Europe. The Iridium satellites produce bright flares as sunlight is reflected from their large flat panels, and may ruin photographs of the sky, unless photography is restricted to the middle of the night, when the satellites are in Earth's shadow.

COMETS

Although there are a number of comets visible at any one time, the majority are very faint, and may be seen only with large telescopes. With the exception of Halley's Comet, which was visible in 1986 and will not return again until 2061, bright comets are unpredictable. We simply do not know when one may appear. Magnificent objects like Comet Hale-Bopp in 1997 appear once or twice a century, and one may have to wait a decade or more to see even a reasonably bright comet like Comet Hyakutake in 1996. If you have a chance to observe a comet, take it – no one can tell when you may see another.

Try not to be disappointed by the appearance of comets. Not many show spectacular tails that are easily visible with the naked eye, like Hale-Bopp and Hyakutake. Many never appear as anything more than a fuzzy patch of light. This is the head or **coma**, and is actually a cloud of material (mostly dust) that the comet has shed into space. Comets consist of a mixture of dust and ice – they are aptly described as 'dirty snowballs' – and as they approach the Sun, they are heated, some of the ice turns to gas, and dust is released.

The parent body, which always remains invisible, is just a few kilometres across (about 17 km for Comet Halley), whereas the coma may be 10,000 km in diameter. (Yet it still appears a tiny speck in the sky!) With powerful telescopes it is sometimes possible to see a bright, star-like **nucleus** in the centre of the coma, but even this is not the actual cometary body, but just the brightest part of the jets of material that are being driven off.

The magnitude that a comet will attain is notoriously difficult to predict. This is because of the nature of comets: some, particularly ones making their first approach to the Sun may be extremely active as large amounts of ice vaporize and stream off into space. Other comets, particularly **periodic comets** – these are defined as comets with periods less than 200 years – may show little activity because their volatile material has been lost over time. As with a comet's appearance, do not expect too much from predictions of a comet's possible brightness. In any case, cometary magnitudes often include the light from the whole of the extended area the outer parts of which are invisible with the naked eye or small instruments.

When close to the Sun, comets may develop a tail or, quite frequently, two tails, both of which point away from the Sun. (On a comet's outward path, after its closest approach to the Sun, the tails always precede it.) The most conspicuous tail is often broad and has a slightly yellowish colour (which is actually reflected sunlight). This is the **dust tail**, which may be enormous and stretch for millions of kilometres. This was the most striking feature of Comet Hale-Bopp. The particles released from comets tend to disperse in the comets' orbital planes, so the way dust tails appear depends greatly on where the particles lie relative to our line of sight. Some tails may be broad fans, others long, curved 'scimitars', and yet others narrow straight spikes. Uneven release of dust may also cause the visible tail to change from night to night.

Comet Hyakutake showed a bright blue gas tail

The other type of tail is the **gas tail**. This is generally almost perfectly straight, and points directly away from the Sun. Emission from the gas causes its colour to be blue, and sometimes streamers within the tail make it appear double or even multiple. This gas tail was quite strong in Hale-Bopp, but was not noticed by many casual observers, who contented themselves with a quick glance. Anyone who allowed their eyes to become dark-adapted (p.10) found it easy enough to see.

Comets originate in an enormous spherical region, halfway to the nearest stars, known as the Oort Cloud (after the astronomer who discovered its existence). Their orbits are randomly orientated in space, so they may approach the inner Solar System from any direction. Unlike the planets, therefore, they are not confined to the zodiacal region, and their paths may take them across any part of the sky. We have been fortunate in recent years, because several comets have passed close to Polaris and were thus circumpolar, remaining visible throughout the night.

The zodiacal light

The dust released from comets not only gives rise to meteors (p.158), but also contributes to a disk of interplanetary dust, the central plane of which roughly coincides with the ecliptic. This dust scatters light from the Sun and gives rise to a phenomenon known as the **zodiacal light**. If your sky is particularly dark and clear in the west around the time of the vernal (spring) equinox, or in the east around the autumnal equinox, you may be lucky enough to see a pale cone of light with its base on the horizon, and pointing up along the ecliptic.

This 'cone' is actually part of an elliptical area of scattered light that is centred on the Sun. It is visible at other times of the year, but is easiest to see around the equinoxes, when the ecliptic makes the greatest angle with the horizon. The dust causing the zodiacal light lies inside the Earth's orbit. If you have exceptionally clear dark skies during the night, you may be able to make out a weak patch of light in the sky, directly opposite the Sun's position. This is the **gegenschein**, again

Comet Hale-Bopp had strong gas and dust trails

caused by light scattered from dust, but this time outside the Earth's orbit.

Both the zodiacal light and the gegenschein are rare, and many experienced astronomers have never seen either. Under truly exceptional conditions, some observers have been lucky enough to see a weak bridge of scattered light (the **zodiacal band**) that joins the zodiacal light to the gegenschein.

STARS AND DEEP-SKY OBJECTS

The colour of stars is often apparent to the naked eye, but becomes more pronounced with any optical aid. Stars' colours are determined by their surface temperatures: blue-white stars such as Rigel are hotter than yellow stars like the Sun, which are themselves hotter than red stars like Betelgeuse. Details of a few prominent stars are listed on p.222.

Stars differ greatly in their sizes and for historical reasons are divided into dwarfs (like the Sun), giants and supergiants.

The Sun is an average-sized star and is about 1,400,000 km in diameter. Giants are some 10 times that size, and the largest supergiants are believed to be as much as 1000 times the size of the Sun. Some giant and supergiant stars are listed on p.220 and p.180. (The stars known as white dwarfs are much smaller, only a few thousand kilometres across. None are readily visible in amateur-sized telescopes.)

Stars frequently occur as a binary system, where two stars are bound together gravitationally and orbit one another. Multiple systems of 3, 4, 5, or even more stars are also known. When two stars happen to lie near the same line of sight, although possibly at different distances, they form a visual double. Some of these double stars are listed on p.204.

Many stars vary in brightness. These variables (some are listed on p.218), change for a whole range of reasons too complicated to explain here. Many, such as δ Cephei (p.194) and Mira (p.196) pulsate, expanding and contracting regularly or irregularly. Others, show sudden explosive flares that are unpredictable. The brightest novae or supernovae may change completely the appearance of a constellation or galaxy.

The distances of stars are often given in light-years. One light-year – a measure of distance, not time – is the distance covered by light (at a speed of about 300,000 km/s) in one year. (It is about 9.5 million million kilometres.) The closest star is about 4.3 light-years away. For comparison, light covers the average distance between the Earth and the Sun (149,597,870 km) in just 8 minutes 19 seconds.

Star clusters

Stars often occur in clusters. Open clusters are irregular concentrations of stars, which usually number from ten to a few hundred. They were all born roughly simultaneously from a cloud of dust and gas. The Pleiades (p.244) is a famous example, and others are listed on p.192.

Globular clusters (p 200) may contain many thousands of stars in a closely-packed spherical region of space. Only the largest telescopes resolve them into individual stars, which are some of the oldest in the Galaxy.

Nebulae

Clouds of gas and dust in interstellar space are known as nebulae. A few glowing gaseous nebulae are listed on p.234. Dust clouds are dark and absorb the light from more distant stars. The most famous dark nebula is the Coalsack, visible only in the southern hemisphere, but the Great Rift in Cygnus (p.202) is produced by the same effect.

Galaxies

Galaxies are giant systems, many thousands of light-years across that contain thousands of millions of stars. Many, such as our own, M31 in Andromeda (p.168), and M33 in Triangulum (p.174) are flattened, with a disk and spiral arms, together with a central bulge. The Sun lies within the disk of the Galaxy, and the band of the Milky Way is the projection of this disk against the sky.

Other galaxies, including some of the largest in the universe, are spherical or elliptical, without a flattened disk. The giant galaxy M87 in Virgo (p.200) is of this type. Some bright galaxies are listed on p.200.

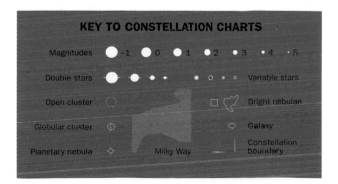

KEY TO CONSTELLATION CHARTS

| Magnitudes | −1 | 0 | 1 | 2 | 3 | 4 | 5 |

| Double stars | | | | | Variable stars |

| Open cluster | | | Bright nebulae |

| Globular cluster | | | Galaxy |

| Planetary nebula | | Milky Way | Constellation boundary |

FACT FILE

SOUTH AT 22:00:
Nov.10

AREA:
(19th) 722 sq. deg.

VARIABLE STARS:
R And

OPEN CLUSTERS:
NGC 752

GALAXIES:
M31, M32, M101

ANDROMEDA
Andromedae • And

Although none of the stars in Andromeda is brighter than magnitude 2, the constellation is fairly easy to recognize. The brightest stars run in a line from α (Alpheratz or Sirrah), at the north-eastern corner of the Great Square of Pegasus, through, δ, β (Mirach) and on to γ (Alamak). In mythology, the constellation represented Andromeda, the beautiful daughter of Cepheus (p.194) and Cassiopeia (p.192), the rulers of Æthiopia. When Cassiopeia boasted of her beauty, compared with the daughters of Poseidon, the enraged god sent Cetus (p.196) to destroy the country. An oracle warned that only the sacrifice of Andromeda would save the kingdom, so she was chained to a rock on the seashore, until rescued by Perseus (p.230). This constellation is therefore sometimes known as The Chained Princess.

Under clear skies, the Great Andromeda Galaxy (M31) is easily visible to the naked eye as a hazy patch. It may be found from Mirach by following the line of the two fainter stars μ and ν. Binoculars show it more clearly but without any detail. The light from this giant spiral galaxy, which is rather larger than our own Milky Way galaxy, has taken about 2.3 million years to reach us. There are two companion galaxies: M32 is sometimes visible in large binoculars under exceptionally good skies, but M110 is visible only with a telescope.

The open cluster NGC 752, almost due south of γ (Alamak), is best observed with binoculars, because it cover quite a large area and is also fairly bright.

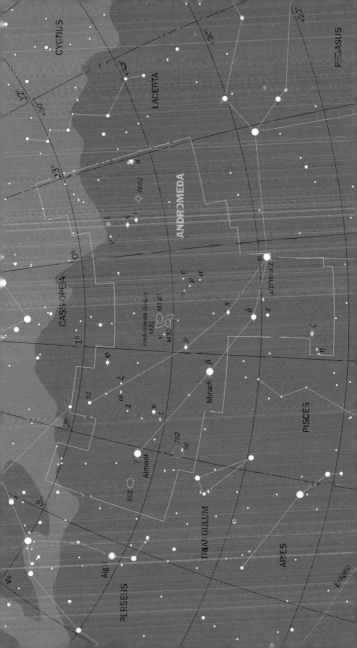

AQUARIUS
Aquarii · Aqr · The Water Bearer

FACT FILE

SOUTH AT 22:00:
Oct.10

AREA:
(10th) 980 sq. deg.

GLOBULAR CLUSTERS:
M2

METEORS:
η Aquarids:
Apr.24-May.20,
maximum May.05

δ Aquarids:
Jly.15-Aug.20,
maximum Jly.28

This zodiacal constellation is extremely ancient, and even in Babylonian times was considered to represent a water-carrier pouring water from a jar (the 'Y'-shaped asterism of γ, η, ζ, and π). The water was imagined to be flowing down to the south-east, towards the bright star Fomalhaut in Piscis Austrinus. It has been suggested that the original association with water was because when β Aquarii was seen in the east just before dawn, it heralded the beginning of the rainy season.

The stars in this constellation may appear relatively faint, but both α (Sadalmelek) and β (Sadalsud) are, in fact, extremely luminous, yellow supergiants. Although their surface temperatures are similar to that of the Sun, each is about 120 times the diameter, and roughly 30,000 times as luminous. They only appear so faint because they are, respectively, about 759 and 612 light-years away.

To the north of Sadalsud lies M2, a moderately bright globular cluster (magnitude 6.5), containing tens of thousands of stars. Its distance is about 40,000 light-years.

Two fine meteor showers have radiants in Aquarius (p.170). The η Aquarids, which peak in May, have a higher rate than the combined rates from the twin streams that make up the δ Aquarids, which reach their maximum in July. Unfortunately neither shower is easy to observe from these latitudes.

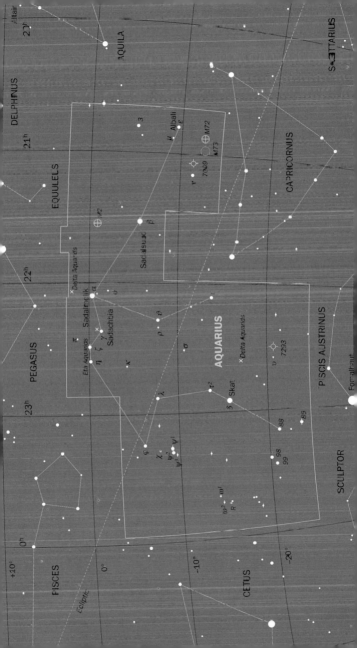

FACT FILE

SOUTH AT 22:00:
Aug.30

AREA:
(22nd) 652

VARIABLE STARS:
η Aql

OPEN CLUSTERS:
NGC 6709,
(M11 in Scutum)

AQUILA
Aquilae · Aql · The Eagle

Aquila is another ancient constellation, originating with the Babylonian astronomers. It later became associated with the Roman god Jove (Jupiter), bearing his thunderbolts in the battle with the Titans. It subsequently carried Ganymede to the heavens, where he became the cupbearer to the gods, and it was then immortalized as the constellation of Aquarius.

Altair, α Aquilae, is one of the three stars that form the prominent Summer Triangle (p.77). It is bright because it is relatively close to us at a distance of about 17 light-years. By contrast, η Aquilae, at a distance of 1173 light-years, is one of the brightest Cepheid variable (p.194) stars known. It has a period of about 7.18 days. At its brightest its magnitude is similar to δ (Deneb Okab), and at its faintest to ι Aquilae.

NGC 6709 is a fairly dense open cluster, visible against the background of the Milky Way, just to the west of the dark dust lane that forms the extension to the Great Rift in Cygnus (p.202). It is probably easiest to locate the open cluster M11 in Scutum (p.240) by star-hopping from λ Aquilae.

FACT FILE
[Ari]

SOUTH AT 22:00:
Nov.20

AREA:
(39th) 441 sq. deg.

FACT FILE
[Tri]

SOUTH AT 22:00:
Nov.20

AREA:
(78th) 132 sq. deg.

GALAXIES:
M33

ARIES
Arietis · Ari · The Ram

TRIANGULUM
Trianguli · Tri · The Triangle

Aries is conventionally regarded as the first constellation in the zodiac. This dates from the time, some 3000 years ago, when the point at which the Sun crossed the celestial equator from south to north at the vernal equinox (p.24) lay within the constellation. Because of precession (p.25) this point, still known as the First Point of Aries, has now shifted into the neighbouring constellation of Pisces. Mythologically, Aries is supposed to be the ram whose fleece turned to gold and was subsequently sought by Jason and the Argonauts.

The other small constellation shown here, Triangulum, is extremely ancient, and was recognized by several different cultures. Quite why these three stars (which are all faint), should have been regarded as forming a constellation, when many other sets of three stars could have been chosen, remains a mystery.

M33 is a spiral galaxy within the Local Group, which includes our Galaxy and M31 in Andromeda. Under exceptional conditions it may be glimpsed with the naked eye – the most distant object (at about 2.7 million light-years) that is visible without instrumental aid. Even with binoculars it remains difficult to see, however, because it is about a degree across (roughly twice the size of the Moon). Its stars are not concentrated in a small area, unlike M31, so its surface brightness is much fainter.

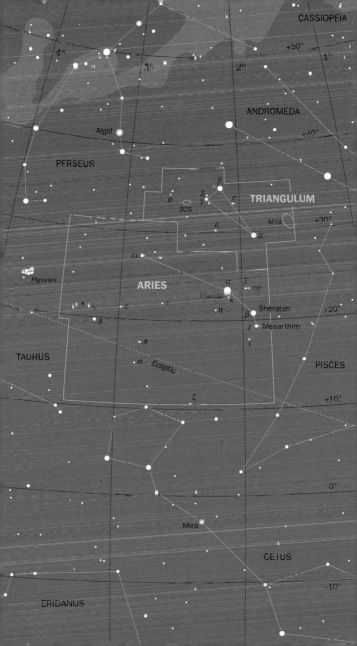

CASSIOPEIA

4ʰ   3ʰ   2ʰ   1ʰ   +50°

ANDROMEDA

Algol   +40°

PERSEUS

β
δ   ε
R   γ
926   TRIANGULUM

δ   M.33   +30°

α

41

Pleiades   ARIES   α   λ
Hamal   k   π
η   κ
Sheratan   +20°
β
ε   ν
γ   Mesarthim
δ   ι

π

TAURUS   σ   Ecliptic

PISCES

ξ   +10°

0°

Mira

CETUS   −10°

ERIDANUS

J	F	M	A	M	J
J	A	S	O	N	D

AURIGA
Aurigae · Aur · The Charioteer

FACT FILE

SOUTH AT 22:00:
Jan.10

AREA:
(21st) 657 sq. deg.

VARIABLE STARS:
ε, ζ

OPEN CLUSTERS: M35,
M36, M37

Although originating in a Babylonian constellation of a chariot, the mythological figure is generally associated with Erichthonius, who like his father, the god Hephaestus (Vulcan) was a cripple, and who invented the chariot so that he could move around. The identification of Capella ('Little She-Goat') as a goat probably arose as a misunderstanding of the Greek word for a stormy wind, which the constellation was thought to herald. The 'Kids' (ε, η, and ζ) were added later as a natural extension of this association. Capella, α Aur, actually consists of a pair of yellow giant stars, orbiting so close to one another that they cannot be separated by even the largest telescopes.

Two of the Kids are remarkable stars. ε Aurigae is an eclipsing binary (p.230) with one of the longest periods known (over 27 years). The last eclipse was in 1983–4. Despite intensive study, the nature of the eclipsing 'star' remains a mystery. It may be a dense disk of dust surrounding a pair of very small, white dwarf stars (p.166). ζ Aurigae is also an eclipsing binary, with a period of 972.16 days. Both stars are supergiants (p.166) but are very different in size. One is about 5 times the size of the Sun, but the other is about 200 times its diameter. Placed in the Solar System it would nearly reach the orbit of the Earth.

Three open clusters are visible in binoculars. M36 and M38 may be found from the star ϑ Aur, with M36 rather smaller and brighter than M38. M37 is larger, almost the size of the Moon and is particularly fine in small telescopes.

J	F	M	A	M	J
J	A	S	O	N	D

FACT FILE

SOUTH AT 22:00:
May.30

AREA:
(13th) 907 sq. deg.

DOUBLE STARS:
μ, ν

VARIABLE STARS:
W

METEORS:
Quadrantids
Jan.1-6, max.
Jan.4, bright
bluish- and
yellowish-white
meteors.

BOÖTES
Boötis · Boo · The Herdsman

In legend, Boötes is supposed to have been accorded a place in the sky for having invented the plough. In another version, he is regarded as holding Canes Venatici (the Hunting Dogs) on the leash, and to be pursuing the two Bears, Ursa Major and Ursa Minor around the sky. The name Arcturus, which was once applied to the whole constellation and not just to α Boötis, also means 'guardian of the bear(s)'.

Arcturus is the brightest star in the northern hemisphere, and the fourth brightest in the whole sky. Because it is bright, its yellowish-orange colour is distinct, and becomes even more apparent when viewed through binoculars.

There are many double stars in Boötes: binoculars reveal μ to be double, and the secondary is itself a binary, but the third component is seen only with larger telescopes. ν Boo consists of a white and orange optical pair that do not form a single system.

W Boötis, near ε, is a red giant variable star. It is semiregular in its behaviour, but although it remains visible in binoculars at all times, it is a bit difficult to study because its changes in brightness are rather small (less than one magnitude) and its colour may cause problems for inexperienced observers.

One of the year's most consistent meteor showers, the Quadrantids, has its radiant in Boötes, near the border with Draco. This shower is named after a former constellation, Quadrans Muralis (the Mural Quadrant), an astronomical instrument that, like the constellation name, is no longer used.

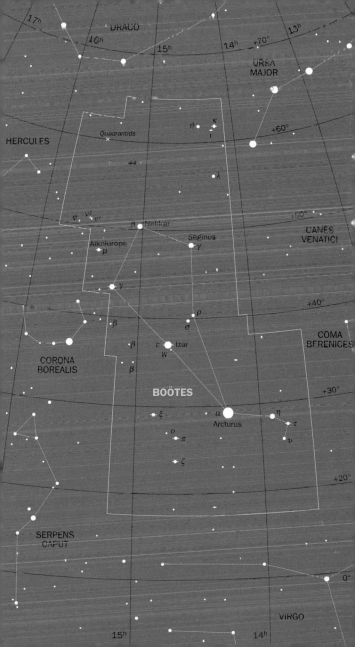

J	F	M	A	M	J
J	A	S	O	N	D

CAMELOPARDALIS
Camelopardalis · Cam · The Giraffe

FACT FILE

SOUTH AT 22:00:
Jan.01

AREA:
(18th) 757 sq. deg.

VARIABLE STARS:
VZ

Camelopardalis is a relatively modern constellation, having been first proposed in 1613 by Petrus Plancius to fill a large, blank area among the northern circumpolar constellations. It was shown in an atlas by Jakob Bartschius in 1624. All its stars are faint (β is actually the brightest, at magnitude 4.0), so any pattern is rather difficult to make out.

α Cam is a brilliant blue-white supergiant star, which is so far away (6940 light-years) that it appears slightly fainter (mag. 4.3) than β. VZ Cam, quite close to the northern border towards Polaris, is an irregular variable star, always visible in binoculars as it changes between about magnitudes 4.8 and 5.2, but its small range makes it rather difficult to study.

Supergiant stars

α Aquarii	β Aquarii	η Aquilae	ζ Aurigae
α Camelopardalis	δ Cephei	α Cygni	μ Cephei
R Coronae Borealis	ζ Geminorum	α Herculis	α Orionis
β Orionis	ε Pegasi	α Scorpii	α Ursae Minoris

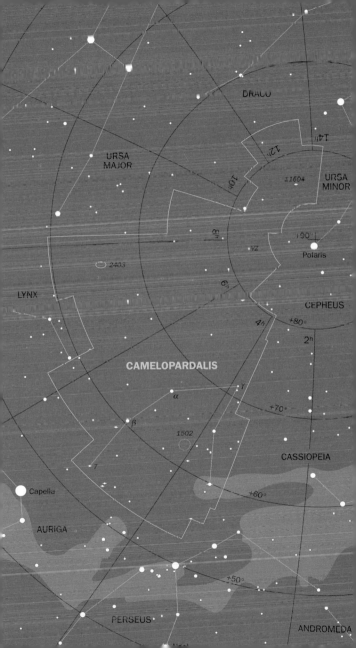

FACT FILE

SOUTH AT 22:00:
Mar.05

AREA:
(31st) 506 sq. deg.

DOUBLE STARS:
ι

OPEN CLUSTERS:
M44, M67

CANCER
Cancri · Cnc · The Crab

Cancer is the least conspicuous of the zodiacal constellations, but 2000 years ago, when it was first described, it was of considerable importance, because the Sun was in the constellation at the summer solstice. Thanks to precession (p.25) the highest point of the ecliptic has moved through Gemini and just crossed into Taurus. A link with the past is preserved in the name of the Tropic of Cancer, where the Sun stands directly overhead at the summer solstice.

The most notable object is the open cluster M44 or Praesepe (the Manger), also known as the Beehive Cluster. It is a large cluster about 1.5° across, visible as a hazy patch to the naked eye, but best viewed with binoculars, which also reveal its brightest member, ε Cancri, at magnitude 6.3. The stars γ and δ Cancri to the immediate north and south, are Asellus Borealis ('northern donkey') and Asellus Australis ('southern donkey') respectively, supposedly feeding at the manger.

M67 is another open cluster, more condensed than M44 and only about 0.5° across. A telescope is required to resolve it into individual stars.

Three of the brightest stars are much larger than the Sun: β, an orange giant (p.166) is actually the brightest star in the constellation at mag. 3.5 δ (Asellus Australis) is slightly smaller (a sub-giant) of mag. 3.9, and ι is a yellow giant of mag. 4.0. It has a mag. 6.6 companion that is just visible with binoculars.

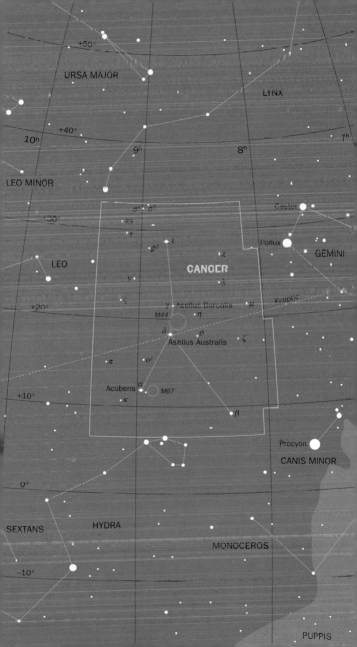

J	F	M	A	M	J
J	A	S	O	N	D

FACT FILE
[CVn]

SOUTH AT 22:00:
May.05

AREA:
(38th) 465 sq. deg.

VARIABLE STARS:
Y

GLOBULAR CLUSTERS:
M3

FACT FILE
[Com]

SOUTH AT 22:00:
May.05

AREA:
(42nd) 386 sq. deg.

CANES VENATICI
Canum Venaticorum · CVn ·
The Hunting Dogs

COMA BERENICES
Comae Berenices · Com ·
Berenices' Hair

These two small constellations lie beneath the tail of Ursa Major and both are of fairly recent origin. Canes Venatici represents two hounds, held on the leash by Boötes, that are pursuing Ursa Major and Ursa Minor. The constellation was proposed by the Polish astronomer Jan Helweke (Hevelius) in 1687. α CVn is sometimes known as Cor Caroli ('Charles' heart') after the executed King Charles I of England.

M3, close to the border with Boötes, is a fine globular cluster that is just at the naked-eye limit, but readily visible in binoculars. It consists of thousands of stars and lies at a distance of 32,200 light-years.

Coma Berenices is also a 'modern' constellation, created by the Belgian cartographer Gerard Mercator in 1551 from part of Leo. It represents a lock of hair that Queen Berenices of Egypt offered in thanks for the safe return of her husband Ptolemy III Euergetes. Most of the scattering of faint stars in this constellation belong to a loose grouping known as the Coma Star Cluster. Both of these constellations contain large numbers of distant galaxies, but these are too faint to see even with binoculars.

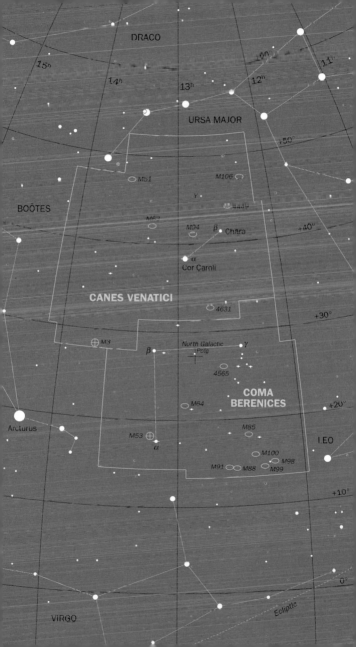

J	F	M	A	M	J
J	A	S	O	N	D

Fact File [CMa]

SOUTH AT 22:00:
Feb.05

AREA:
(43rd) 380 sq. deg.

OPEN CLUSTERS:
M41, NGC 2362

Fact File [Lep]

SOUTH AT 22:00:
Jan.15

AREA:
(51st) 290 sq. deg.

VARIABLE STARS:
R

CANIS MAJOR
Canis Majoris • CMa •
The Greater Dog

LEPUS
Leporis • Lep • The Hare

Canis Major is dominated by Sirius, the brightest star in the sky. It was of great significance to the ancient Egyptians, because when it rose just before the Sun at dawn in the late summer it heralded the annual flooding of the Nile, which covered the fields with fertile silt. This 'heliacal rising' marked the beginning of their calendar year.

Sirius is accompanied by Sirius B, the first white dwarf star to be discovered, whose material is so dense that, if it could exist on Earth, a teaspoonful would weigh five tonnes. Unfortunately it is visible only in the largest amateur telescopes.

About 4° south of Sirius is M41, a beautiful, fairly compact open cluster, with about 50 stars in an area about the same as that of the Full Moon. Farther south another open cluster, NGC 2362, appears to surround bluish-white τ Canis Majoris. In fact, τ is a foreground star at a distance of about 3,200 light-years, whereas the distance of NGC 2362 is 5,000 light-years.

This small constellation was once said to represent Orion's chair, but because he was a hunter, it later seemed more appropriate for it to be seen as a hare, lying at his feet. Its most notable star is the variable, R Leporis, which becomes visible to the naked eye at maximum. Also known as Hind's Crimson Star (after the British astronomer who described it), it is one of the reddest stars in the sky. Its colour is more distinct through binoculars.

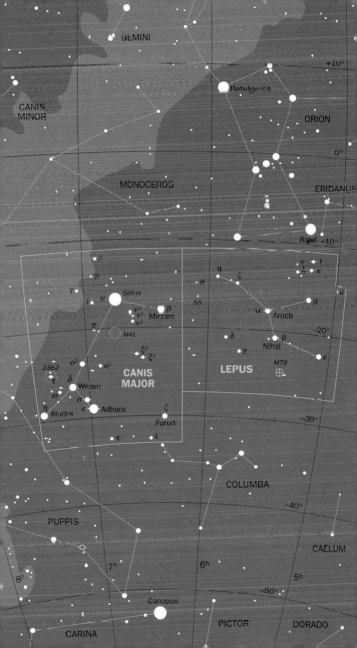

Fact File [CMi]

SOUTH AT 22:00:
Feb.15

AREA:
(71st) 183 sq.deg.

Fact File [Mon]

SOUTH AT 22:00:
Feb.05

AREA:
(35th) 482 sq.deg.

DOUBLE STARS:
δ

OPEN CLUSTERS:
M50, NGC 2232,
NGC 2264, NGC
2301

CANIS MINOR
Canis Minoris · CMi ·
The Lesser Dog

MONOCEROS
Monocerotis · Mon · The Unicorn

This tiny constellation has only two moderately bright stars, α Procyon – whose name, meaning 'before the dog', is derived from the fact that it rises just before Sirius – and Gomeisa. By strange coincidence, Procyon, like Sirius, is accompanied by a white-dwarf companion, visible only with professional-size instruments.

Monoceros is a faint constellation that (like Camelopardalis) was first introduced in 1613 by Petrus Plancius. It lies largely within the 'Winter Triangle' – the three stars Betelgeuse, Procyon and Sirius – against the backdrop of the Milky Way. It contains numerous clusters and various nebulae, although the latter are too faint for easy visibility.

δ Mon (mag. 4.2) forms a wide double with 21 Mon (mag. 5.5). M50 is an open cluster that contains about 80 stars, 2900 light-years away. NGC 2232 is smaller, with about 20 stars around bluish-white 10 Mon (mag. 5.1). NGC 2264 has about 40 members, and includes the brilliant low-amplitude variable S Mon (mag. 4.7). Finally, NGC 2301 is yet another open cluster with about 80 stars. All are visible in binoculars.

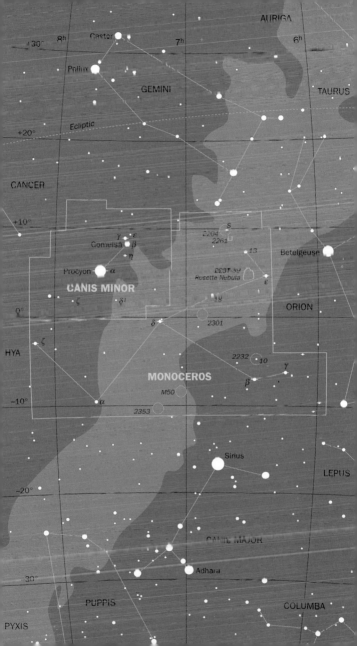

FACT FILE

SOUTH AT 22:00:
Sep.01

AREA:·
(40th) 414 sq.deg.

DOUBLE STARS:
α

CAPRICORNUS
Capricorni • Cap • The Sea Goat

Capricornus is an ancient constellation that has always been associated with a goat. It is commonly depicted as a goat with the tail of a fish, and this may relate to the legend that when the god Pan jumped into the Nile to avoid the monster Typhon, the part underwater turned into a fish, while that above the water remained a goat.

Just as Cancer (p.182) was associated with the summer solstice, the Sun was once in Capricornus at the winter solstice. (It now lies in Sagittarius.) The link with the past is preserved in the Tropic of Capricorn, the southern latitude at which the Sun is overhead at the (northern) winter solstice.

On a clear night it is possible to see with the naked eye that α Capricorni is actually double, and the stars are often marked α^1 (Prima Giedi) and α^2 (Secunda Giedi) on charts. Although these merely happen to lie close to the same line of sight and are really at distances of approximately 690 and 110 light-years, respectively, both are actually true binary stars.

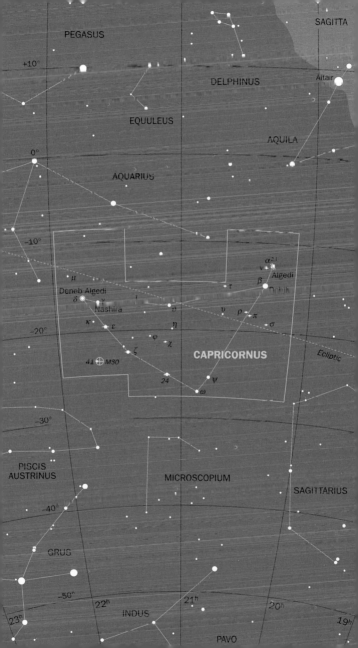

J	F	M	A	M	J
J	A	S	O	N	D

FACT FILE

SOUTH AT 22:00:
Nov.05

AREA:
(25th) 598 sq.deg.

VARIABLE STARS:
γ

OPEN CLUSTERS:
M52, NGC 663

CASSIOPEIA
Cassiopeiae • Cas

In legend, Cassiopeia was Queen of Æthiopia (Joppa in ancient Palestine), wife of King Cepheus and the mother of Andromeda (p.168). She is depicted as sitting on her throne and one embellishment to the tale is that she was chained to it as punishment for her boastfulness, and thus has to suffer the daily indignity of hanging upside down in the sky.

γ Cas is an unusual variable star (known as a shell star) that fluctuates irregularly in brightness. It is normally about mag. 2.5, but has varied between mag. 1.6 and 3.0. It is rotating so rapidly that it sheds material into space from its equatorial regions. The star tends to brighten when it ejects a large shell of gas.

Cassiopeia lies in the Milky Way and therefore contains many open clusters. Try sweeping the whole area with binoculars. The brightest is M52, which contains about 100 stars, but it is not very striking in binoculars. NGC 663 is fairly compact, but the stars are quite faint.

Some interesting open clusters

M36 (Auriga)	M37 (Auriga)	M44 (Cancer)
M41 (Canis Major)	M35 (Gemini)	M48 (Hydra)
h & χ Persei	M24 (Sagittarius)	M6 (Scorpius)
M7 (Scorpius)	M11 (Scutum)	M45 (Taurus)

J	F	M	A	M	J
J	A	S	O	N	D

FACT FILE

SOUTH AT 22:00:
Sep.20

AREA:
588 sq.deg. (27th)

DOUBLE STARS:
δ

VARIABLE STARS:
δ, μ

OPEN CLUSTERS:
NGC 7160, IC 1396

CEPHEUS
Cephei · Cep

This rather faint constellation represents the legendary King Cepheus of ancient Æthiopia, husband of Cassiopeia (p.192) and father of Andromeda (p.168).

Two stars are of particular interest. δ Cephei is the prototype of an important class of variables, known as Cepheid variables. These stars fluctuate in brightness extremely regularly. The periods (generally a few days) are related to the stars' true (rather than apparent) luminosity. By measuring a Cepheid's period, it is possible to determine its actual luminosity, and from that and its apparent brightness, work out its distance. Cepheid stars are thus 'standard candles' for determining distances in space.

For δ Cep, its period is 5.366341 days, and it varies between the brightness of ζ Cep (mag. 3.5) and ε Cep (mag. 4.4). Its surface temperature is similar to that of the Sun (about 6000°C), but it is a yellow supergiant, lying at a distance of 982 light-years.

The second interesting star is μ Cep. Because of its striking colour (particularly noticeable with binoculars), William Herschel called it the Garnet Star. It is the largest star currently known, 2400 times the diameter of the Sun. If placed in the Solar System, it would engulf all the planets out to (and including) Saturn. It is a semiregular variable that changes between magnitudes 3.4 and 5.1 with periods of around 730 and 4400 days, together with intermittent periods of low activity.

CETUS

Ceti · Cet · The Sea Monster

FACT FILE

SOUTH AT 22:00:
Nov.20

AREA:
(4th) 1231 sq.deg.

DOUBLE STARS:
α

VARIABLE STARS:
o

In mythology, Cetus represents the monster sent to ravage the coast of Æthiopia, and about to attack Andromeda (p.168), before being turned to stone when Perseus (p.230) showed it the head of Medusa. Menkar, α Cet, is a binocular double, consisting of a red giant (mag. 2.5) and a bluish-white star (mag. 5.6).

The most important object in this constellation is probably the famous star, o Ceti (Mira – 'the Wonderful'), the first variable star to be recognized. Discovered in 1596 by the Friesian astronomer Fabricius, it was not until 1638 that Holwarda realized that its brightness changes regularly, with a period about 330 days. It is now taken as the prototype of the class of long-period variables, red giant stars that expand and contract in a more-or-less regular manner. Many thousands of these are now known.

Mira varies from about mag. 3.4 at maximum, when it is easily visible with the naked eye, to around 9.5 at minimum, when it is beyond the grasp of most binoculars. At times it may exceed this range and be even brighter or fainter.

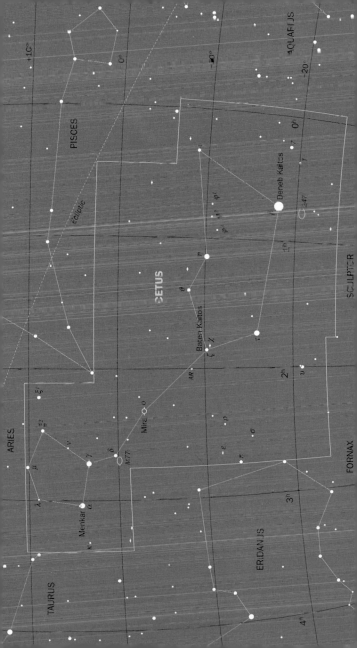

| J | F | M | A | M | J |
| J | A | S | O | N | D |

FACT FILE

SOUTH AT 22:00:
Jun.15

AREA:
(73rd) 179 sq.deg.

DOUBLE STARS:
ν

VARIABLE STARS:
R

CORONA BOREALIS
Coronae Borealis · CrB ·
The Northern Crown

This small, but distinctive semicircular constellation is easily identified just to the east of Boötes. In legend, it represents the crown of Ariadne who, after being deserted by Theseus, became the wife of the god Dionysus. She was granted immortality by Zeus, who placed her bridal gift, a crown, among the stars.

Of the arc of seven stars, six are magnitude 4, and the seventh, α CrB, Gemma or Alphecca, is brighter, at magnitude 2.2. Away from the main body of the constellation, ν consists of a wide double, with two orange giant stars.

The most notable star is R CrB, the prototype of a small, and very unusual class of supergiant variable stars. Normally at about magnitude 5.9, it suddenly, and completely unpredictably, drops to fainter magnitudes over a period of days or weeks. The depths of the fades differ greatly, but may sometimes be as low as mag. 14–15. These fades occur when a cloud of carbon particles condenses in the star's outer atmosphere, and blocks most of the visible light. The cloud gradually disperses, but it may take months or well over a year for the star to regain its normal brightness. Only about two dozen such stars are known. Spotting the onset of such an event is of great interest to professional astronomers, so the star is regularly monitored by amateurs using binoculars.

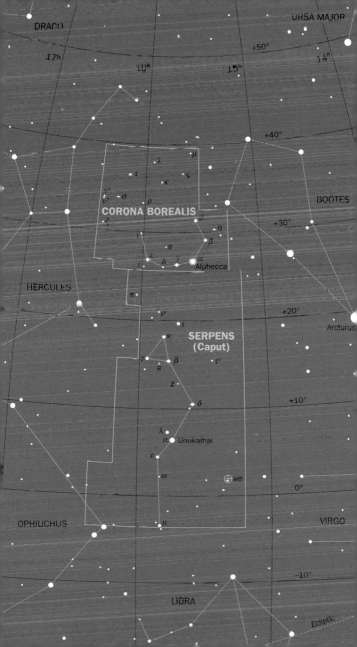

Fact File [Cor]
SOUTH AT 22:00:
Apr.25
AREA:
(70th) 184 sq. deg.

Fact File [Crt]
SOUTH AT 22:00:
Apr.10
AREA:
(53rd) 282 sq. deg.

CORVUS
Corvi · Crv · The Crow

CRATER
Crateris · Crt · The Cup

The two small constellations of Corvus and
Crater lie between Virgo and Hydra. Both are
faint (particularly Crater) and do not contain
any notable objects visible with the naked eye
or binoculars.

Some bright globular clusters

M2 (Aquarius)	M3 (Canes Venatici)	M13 (Hercules)
M92 (Hercules)	M9 (Ophiuchus)	M15 (Pegasus)
M5 (Serpens)	M22 (Sagittarius)	M4 (Scorpius)

Some bright galaxies

M31 (Andromeda)	M65 (Leo)	M66 (Leo)
M33 (Triangulum)	M81 (Ursa Major)	M48 (Virgo)
M84 (Virgo)	M86 (Virgo)	M87 (Virgo)

FACT FILE

SOUTH AT 22:00:
Aug.25

AREA:
804 sq. deg. (16th)

DOUBLE STARS:
O¹

VARIABLE STARS:
χ

OPEN CLUSTERS:
M39

NEBULAE:
NGC 7000

FACT FILE [Lac]

SOUTH AT 22:00:
Sep.25

AREA:
(68th) 201 sq. deg.

CYGNUS
Cygni · Cyg · The Swan

LACERTA
Lacertae · Lac · The Lizard

Cygnus is one of the finest constellations in the sky and lies in a striking part of the Milky Way. It contains many open clusters, and one of the brightest is M39, north of ρ Cygni. The gaseous North America Nebula (NGC 7000), near Deneb, is just visible to the naked eye. It is well worth sweeping the whole constellation with low-power binoculars (or opera glasses).

Deneb, α Cygni, is a remarkable, brilliant blue-white supergiant, about 160,000 times as luminous as the Sun. It lies almost 3230 light-years away, much farther than the other two stars of the Summer Triangle, Vega and Altair, which are at roughly 25 and 17 light-years, respectively.

Under clear skies you should be able to see the Great Rift: a dark patch that runs down the centre of the Milky Way. This is caused by clouds of dust that obscure more distant stars.

O¹ Cyg is a beautiful double, with a pair of orange and bluish stars. You may be able to glimpse another fainter blue star close to the orange component.

χ Cygni is a variable with an extremely large amplitude of about 10 magnitudes (which means that its luminosity changes by a factor of 10,000). Its period is about 408 days. At maximum it is easily visible to the naked eye at magnitude 4 to 5. At minimum it sinks to mag. 13 or 14.

Lacerta is a small constellation that consists of a zigzag of stars between Cygnus and Cassiopeia. It was originally proposed by Hevelius, the famous astronomer from Danzig, but contains no objects bright enough to be seen with the naked eye or binoculars.

J	F	M	A	M	J
J	A	S	O	N	D

DELPHINUS
Delphini · Del · The Dolphin

EQUULEUS
Equulei · Equ · The Little Horse

FACT FILE [Del]

SOUTH AT 22:00:
Sep.01

AREA:
(69th) 189 sq. deg.

FACT FILE [Equ]

SOUTH AT 22:00:
Sep.10

AREA:
72 sq. deg. (87th)

DOUBLE STARS:
γ

Delphinus is a small, but very distinct constellation. It is said to represent the dolphin that rescued the poet and musician Arion from drowning.

Equuleus is the second-smallest constellation (after Crux, the Southern Cross), and contains little of interest, although γ Equ, a yellow star of mag. 4.7, is an optical double with 6 Equ, a white star at mag. 6.1.

Double stars

μ Boötis	ν Boötis	ι Cancri	δ Monocerotis
α Capricorni	ν Coronae Borealis	β Cygni	o¹ Cygni
γ Equulei	ν Draconis	ν Geminorum	ζ Geminorum
α Leonis	γ Leonis	ζ Leonis	ι Librae
δ Lyrae	ε Lyrae	ρ Ophiuchi	δ Orionis
σ Orionis	ε Pegasi	π Pegasi	κ Piscium
ρ Piscium	ω Scorpii	κ Tauri	ϑ Tauri
σ Tauri	ζ Ursae Majoris	γ Ursae Minoris	α Vulpeculae

J	F	M	A	M	J
J	A	S	O	N	D

FACT FILE

CIRCUMPOLAR

AREA:
(8th) 1083 sq. deg.

DOUBLE STARS:
ν, 16/17, 39

DRACO

Draconis · Dra · The Dragon

Draco is an ancient constellation, despite being relatively inconspicuous, with no stars brighter than magnitude 2. About 5000 years ago, at the time when the Egyptian were building the Pyramids, α Dra, Thuban, was the pole star. It has since lost that distinction because of precession (p.25). The quadrilateral 'head' just fills the field of view of most 7x binoculars.

Binoculars reveal that ν is a fine double, consisting of twin white stars of mag. 4.9. Flamsteed 16 & 17 are bluish-white stars of mag. 5.1 and 5.5 that form a wide double. Flamsteed 39 is a similar system, with yellow and blue components of mag. 5.0 and 7.4.

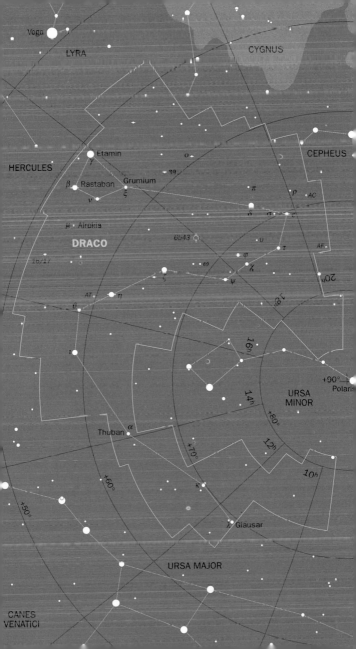

| J | F | M | A | M | J |
| J | A | S | O | N | D |

ERIDANUS
Eridani · Eri · The River

FACT FILE

SOUTH AT 22:00:
Dec.20

AREA:
(6th) 1138 sq. deg.

Eridanus is a large constellation but one that few people recognize, mainly because most of its stars are fairly faint. It starts at γ, right next to Rigel in Orion, makes a wide sweep to the west and then trends south-west until ending at bright α Eri, Achernar (Arabic for 'river's end'), in the far south, permanently invisible to anyone farther north than about latitude 30°N. At one time ϑ Eri (mag. 2.9) was regarded as being the end of the river and called Achernar. When the constellation was extended to its current length, ϑ Eri was renamed Acamar.

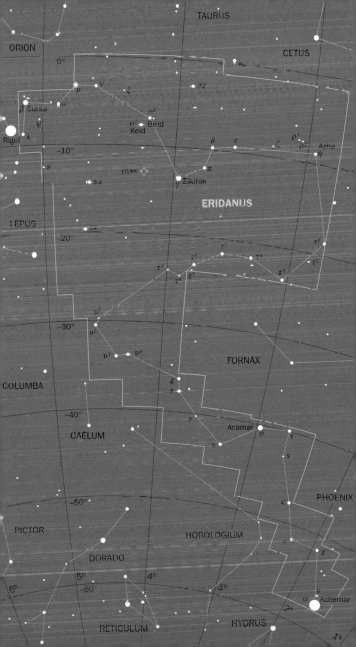

FACT FILE

SOUTH AT 22:00:
Feb.05

AREA:
(30th) 514 sq. deg.

DOUBLE STARS:
ζ, ν

VARIABLE STARS:
ζ, ν

OPEN CLUSTERS:
M35

METEORS:
Geminids

GEMINI
Geminorum · Gem · The Twins

It is a strange coincidence that many of the brightest stars in this constellation of the Twins are double or multiple. Unfortunately few are visible in binoculars. Castor itself, α Gem, is a remarkable object, because it consists of no fewer than six components, in three pairs, all orbiting one another in a single complex system.

ν Gem is a wide double with a bluish-white giant star of mag. 4.1, and a mag. 8.7 companion.

ζ Gem is a binocular double, consisting of a yellow supergiant and an unrelated star of mag. 7.6. The supergiant is a Cepheid variable (p.194) which varies between 3.7 and 4.2, with a period of 10.2 days. This small range is at the limit of detectability with the naked eye, except for experienced observers.

η Gem is a red giant star that varies in a semiregular fashion between 3.2 and 3.9 and displays an approximate period of about 233 days. There is a close companion, but it requires a moderate-sized telescope to be visible.

M35 is a fine open cluster that covers roughly the same area as the Moon. It is visible to the naked eye and through binoculars. It contains about 200 stars and is at a distance of 2,800 light-years.

J	F	M	A	M	J
J	A	S	O	N	D

HERCULES
Herculis • Her

FACT FILE

SOUTH AT 22:00:
Jly.05

AREA:
(5th) 1225 sq.deg.

GLOBULAR CLUSTERS:
M13, M92

Hercules is a sprawling constellation that mythological representations show as the famous hero, kneeling with one foot on the head of Draco. In earlier times, he appeared the right way up, but now, because of precession (p.25), appears more-or-less upside-down. The star representing the head, α Her, Ras Algethi (Arabic for 'kneeler's head') is at the southern boundary. It is a red star, on the borderline between giants and supergiants, and is probably about 500 times the size of the Sun. It varies in brightness by about 1 magnitude, but as with most red stars, it is difficult for inexperienced observers to estimate its brightness accurately.

Hercules contains two notable globular clusters. M13 is the brightest globular in the northern sky and is visible with the naked eye under good conditions. It contains around 300,000 stars and lies at a distance of 23,300 light-years. M92 is slightly farther away, at 25,400 light-years, and fainter, but is still easily seen with binoculars, especially because it is rather more condensed at the centre.

HYDRA

Hydra • Hya • The Water Snake

FACT FILE

SOUTH AT 22:00:
Apr.05 (head)

AREA:
(1st) 1303 sq. deg.

VARIABLE STARS:
R

OPEN CLUSTERS:
M48

GALAXIES:
M83

Hydra is the largest constellation in the sky, but the majority of its stars are faint, so it is not particularly distinctive. Apart from α Hya, Alphard (Arabic for 'the solitary one') an orange giant of mag. 2.0, the most notable feature is the 'head' of Hydra, an attractive asterism consisting of six stars.

On the western border with Monoceros lies M48, just detectable with the naked eye, but seen as a large, somewhat triangular, cluster of about 80 stars in binoculars.

R Hya was the fourth variable star to be discovered (in 1704). It is a red giant very similar to Mira (p.196). Despite its distance (2,013 light-years) it may reach magnitude 3.5 at maximum. Its minimum magnitude is about 10.9, and its period is 389 days.

U Hya is a very cool, red giant variable that fluctuates irregularly between magnitudes 4.3 and 6.6. It is thus always visible in binoculars but its deep red colour makes it difficult to estimate accurately.

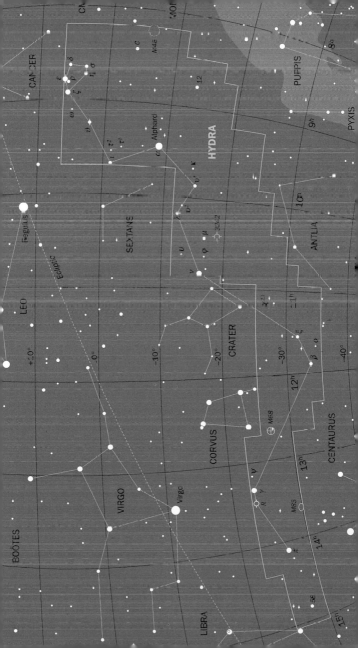

J	F	M	A	M	J
J	A	S	O	N	D

FACT FILE [Leo]

SOUTH AT 22:00:
Apr.01

AREA:
(12th) 947 sq. deg.

DOUBLE STARS:
α, γ, ζ

VARIABLE STARS:
R

GALAXIES:
M65, M66

METEORS:
Leonids

FACT FILE [LMi]

SOUTH AT 22:00:
Apr.01

AREA:
(64th) 232 sq. deg.

LEO
Leonis · Leo · The Lion

LEO MINOR
Leonis Minoris · LMi ·
The Little Lion

Leo is an ancient constellation, described by the Sumerians and Babylonians, and probably associated with the Sun, which then lay in the constellation at the summer solstice. The Greeks saw it as the mythical Nemean Lion, slain by Hercules as one of his twelve labours.

Regulus, α Leo ('little lion') at mag. 1.4, forms a wide double with a 7.7 mag star. γ Leo (Algieba: 'the forehead') is accompanied by a yellow star, 40 Leonis, of mag. 4.8. Telescopes show Algieba itself as a fine binary. Binoculars reveal ζ Leo as a triple system: a white giant, mag. 3.4, a yellowish white star, mag. 5.0, and a yellow star of mag. 6.0.

R Leonis is a well-known variable, close to Regulus. Similar to Mira (p.196) it ranges between magnitudes 5.9 and 11, with a period of 310 days.

The spiral galaxies M65 and M66 are just visible in binoculars, but appear only as featureless patches on the sky.

Spectacular Leonid meteor storms occur at 33-year intervals, most notably in 1966, when the estimated rate reached 140,000 per hour for a short period. This incredible display was visible over a small part of North America. At the time of writing it is impossible to predict if a similar event will occur in 1999.

The tiny constellation of Leo Minor was introduced by Hevelius in 1687. It has been ignored by most subsequent astronomers, because there are few objects of interest. Even the brightest star is labelled 'β' rather than 'α'.

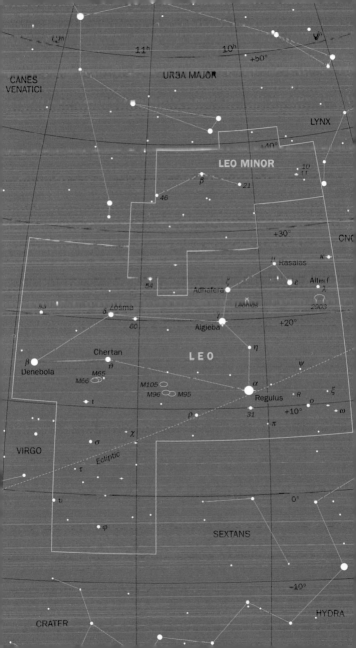

FACT FILE

SOUTH AT 22:00:
Jun.10

AREA:
(29th) 538 sq. deg.

DOUBLE STARS:
α

VARIABLE STARS:
δ

LIBRA
Librae · Lib · The Balance

This constellation once formed part of Scorpio to the east, but was separated by the Romans into a separate zodiacal constellation. The association with Scorpio persists in the Arabic names of its brightest stars. α Lib (Zubenelgenubi – 'southern claw') is a wide double: $α^1$, a pale yellow star of mag. 5.2, and brighter $α^2$, a white star of mag. 2.8. β Lib (Zubeneschamali – 'northern claw') is a rarity: one of the few stars to appear slightly greenish. γ Lib is known as Zubenelakrab ('scorpion's claw').

ι Lib (mag. 4.5) forms an apparent double with nearby 25 Lib (mag. 6.1), which is actually well over 100 light-years closer to us. δ Lib is a Cepheid variable (p.194) ranging from mag. 4.9–5.9, with a period of 2.3 days.

Some bright variable stars

η Aquilae	ε Aurigae	ζ Aurigae	W Boötis
γ Cassiopeiae	δ Cephei	μ Cephei	o Ceti
R Coronae Borealis	χ Cygni	ζ Geminorum	α Herculis
R Leporis	δ Librae	β Lyrae	α Orionis
β Persei	λ Tauri		

J	F	M	A	M	J
J	A	S	O	N	D

LYNX
Lyncis · Lyn · The Lynx

FACT FILE

SOUTH AT 22:00:
Feb.20

AREA:
(28th) 545 sq. deg.

This is an extremely faint constellation, and it has been said that when it was introduced by Hevelius in 1687 he gave it this name, because you need to have the eyes of a lynx to see it. (Hevelius himself was noted for his keen eyesight.) The brightest star, α Lyn (mag. 3.1), is a red giant at a distance of 222 light-years, and has recently been found to be slightly variable.

Giant stars

α Aurigae	W Boötis	β Cancri	ι Cancri
α Ceti	o Ceti	ν Coronae Borealis	η Geminorum
ν Geminorum	α Hydrae	R Hydrae	U Hydrae
ζ Leonis	α Lyncis	δ² Lyrae	π Pegasi
94 Piscium	α Tauri	γ Ursae Minoris	α Vulpeculae

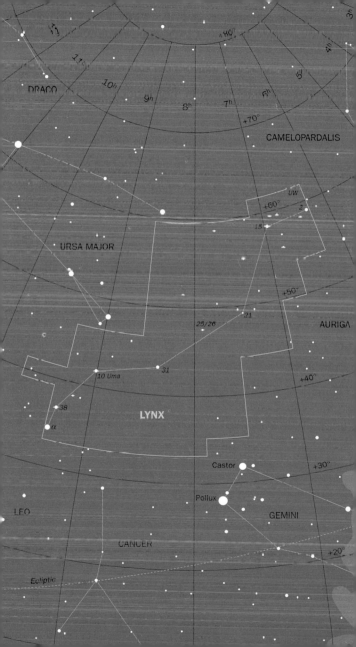

J	F	M	A	M	J
J	A	S	O	N	D

FACT FILE

SOUTH AT 22:00:
Aug.01

AREA:
(52nd) 286 sq. deg.

DOUBLE STARS:
δ, ε, ζ

VARIABLE STARS:
β

METEORS:
Lyrids, Apr.19-25,
max. Apr.22.

LYRA
Lyrae • Lyr • The Lyre

Although small, this constellation contains Vega, a brilliant, white mag. 0.03 star, one of the Summer Triangle with Deneb in Cygnus and Altair in Aquila. The constellation itself is supposed to represent either the lyre belonging to Orpheus, or that of Arion (p.204) who was rescued from drowning by a dolphin.

β Lyrae is a famous variable star that ranges between mags. 3.3 and 4.3 every 12.9 days. It consists of two stars that orbit so closely that they are distorted from true spheres. They regularly eclipse one another causing the changes in brightness.

People with keen eyesight are able to see that δ Lyr is an apparent double, consisting of $δ^1$, a bluish-white mag. 5.6 star 1,080 light-years away, and $δ^2$, a red giant of mag. 4.2, 900 light-years distant.

ε Lyr is the famous 'Double Double'. Binoculars show it as two stars, $ε^1$ and $ε^2$ with magnitudes of 4.7 and 4.6, respectively. Telescopes reveal that both stars are themselves binaries, so this is actually a quadruple system. ζ Lyr is a wide double, mags. 4.4 and 5.7.

Star colours and temperatures

Rigel	Blue-white	11,550
Vega	White	9,960
Sun	Yellow	5,800
Arcturus	Orange	4,420
Betelgeuse	Red	3,450

FACT FILE [Oph]

SOUTH AT 22:00:
Jly.10

AREA:
(11th) 948 sq. deg.

DOUBLE STARS:
ρ

OPEN CLUSTERS:
NGC 6633, IC 4665

GLOBULAR CLUSTERS:
M9, M10, M12,
M14, M19, M62

FACT FILE [Ser]

SOUTH AT 22:00:
Jly.05 (Caput),
Aug.10 (Cauda)

AREA:
(23rd) 637 sq. deg.

DOUBLE STARS:
β

GLOBULAR CLUSTERS:
M5

OPHIUCHUS
Ophiuchi · Oph · The Serpent Bearer

SERPENS
Serpentis · Ser · The Serpent

Ophiuchus is an ancient constellation, usually considered to represent Aesculapius, the legendary healer. (The serpent that he carries, entwined about a staff, has persisted to the current day as the symbol of the medical profession.) The constellation also contains a large portion of the ecliptic, but despite this is not often regarded as a zodiacal constellation. Serpens is the only constellation that is in two halves: Serpens Caput (in the west), the head, and Serpens Cauda (in the east), the tail.

Ophiuchus is close to the centre of the Milky Way (in Sagittarius) and contains numerous globular clusters. M9, M10, M12, M14, M19, and M62 are all visible in binoculars, as is M5 in Serpens, which many regard as second only to M13 in Hercules.

Two open clusters, both containing about 30 stars, are found in the north-eastern part of Ophiuchus: NGC 6633 and IC 4665.

ρ Oph is a triple system in binoculars: a mag. 5.0 star (actually double, but requiring a telescope to be visible as such), and widely separated mag. 6.7 and 7.3 companions. β Ser (mag. 3.7) is a wide double with a star of mag. 6.7 to the north.

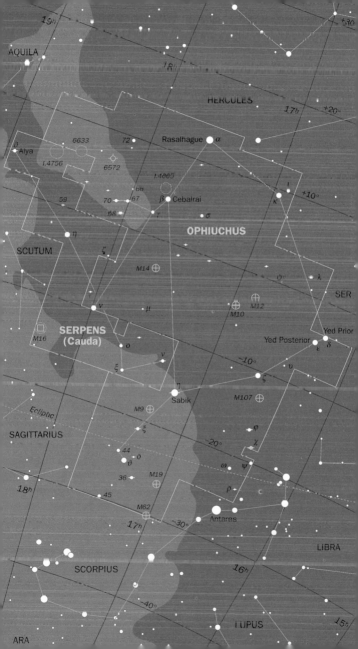

ORION
Orionis • Ori

FACT FILE

SOUTH AT 22:00:
Jan.30

AREA:
(26th) 594 sq. deg.

DOUBLE STARS:
δ

VARIABLE STARS:
α

NEBULAE:
M42

METEORS:
Orionids

Orion is a spectacular constellation, although its most famous object, the Great Orion Nebula, M42, merely appears as a hazy patch to the naked eye. Under good conditions it may show some hints of structure with binoculars. This vast cloud of dust and gas is a gigantic stellar nursery and contains numerous young stars. To the eye, the gas glows green, but even short-exposure photographs (p.16) show the nebula as a distinctive shade of pink that stands out against the variously coloured stars.

Betelgeuse, α Ori, is an enormous, red supergiant star about 800 times the diameter of the Sun. It is also variable, ranging from about mag. 0.3 to 1.2, and sometimes shows a periodicity of about 2335 days (roughly seven years). Rigel, β Ori, is a brilliant, blue-white supergiant, approximately 50,000 times as luminous as the Sun. Alnilam, ε Ori, is a very similar star, but appears fainter because it is farther away.

δ Ori (Mintaka) lies slightly below the celestial equator. It is a wide double with a bright, bluish-white star of mag. 2.3, and an unrelated star of mag. 6.8. σ Ori is actually a multiple system, but binoculars reveal just the primary, mag. 3.8, and one companion of mag. 6.7.

PEGASUS

Pegasi • Peg

FACT FILE

SOUTH AT 22:00:
Oct.20

AREA:
(7th) 1121 sq. deg.

DOUBLE STARS:
ε, π

GLOBULAR CLUSTERS:
M15

Although one of the stars actually forms part of Andromeda, the Great Square of Pegasus is an easily recognized landmark in the sky. This is partly because there are remarkably few stars in this area, despite Pegasus being a large constellation. One of the tests of seeing conditions is to count the number of stars visible within the Great Square. Under really fine skies one may be able to see 12 or 13.

Enif, ε Peg, representing the nose of Pegasus, is a yellow supergiant star, mag. 2.4. With high-quality binoculars you may be able to see a wide companion, mag. 8.4. π Peg, close to the border with Lacerta, is a wide double: π^1 is a yellow giant, mag. 5.6, and π^2 a white giant, mag. 4.3.

As if to make up for the lack of stars, M15 is an outstanding globular cluster not far from Enif. It may just be glimpsed with the naked eye, but is best viewed with binoculars. It lies at a great distance, 30,600 light-years.

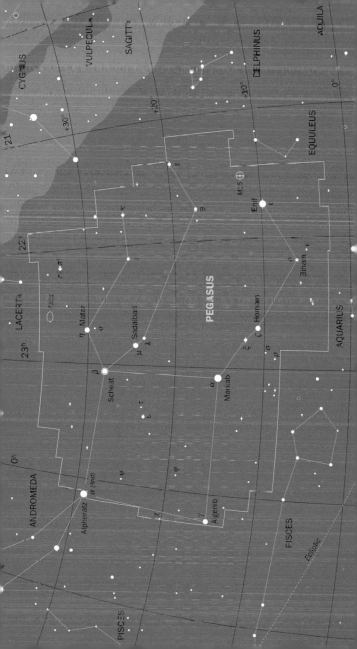

J	F	M	A	M	J
J	A	S	O	N	D

FACT FILE

SOUTH AT 22:00:
Dec.30

AREA:
(24th) 615 sq. deg.

VARIABLE STARS:
β

OPEN CLUSTERS:
h, χ

METEORS:
Perseids

PERSEUS
Persei · Per

In legend, Perseus slew the dreaded Medusa and used her head to turn Cetus to stone, thus saving Andromeda (p.168). In pictorial representations he is shown carrying the severed head of Medusa, marked by the star Algol (Arabic for 'the Demon'), β Per. This star is a famous eclipsing binary. Every 2.87 days the fainter star passes in front of the brighter component, and the combined magnitude fades from mag. 2.1 to 3.4, in an eclipse that lasts about 10 hours.

ρ Per, to the south of Algol, is another variable, this time semiregular, varying between mag. 3.3 and 4.0 and occasionally exhibiting a period of about 50 days.

The other famous sight in Perseus is the Double Cluster, h and χ Persei, which may be found by following the upper 'arm' towards Cassiopeia. Also known as NGC 869 and NGC 884, these clusters are visible to the naked eye and each covers an area about the size of the Moon. In binoculars the whole area is rich in stars. NGC 869 (the one closer to Cassiopeia) is brighter, and contains about 200 stars. NGC 884 has about 150 members. Both are relatively young clusters (on astronomical time-scales): NGC 869 is about 6 million years old, and NGC 884, 3 million. They lie at 7,500 and 7,100 light-years distance, respectively.

PISCES

Piscium • Psc • The Fish

Pictorial representations of this constellation show it as two fish, each with a ribbon tied to its tail, and the ribbons themselves joined with a knot, marked by Alrescha (Arabic 'the cord'), α Psc, in the far east of the constellation. The western fish is marked by a ring of stars, known as 'The Circlet' below the Great Square of Pegasus. The eastern fish, below Andromeda, is not as distinct.

The vernal equinox (p.24) now lies in Pisces, and the point at which the ecliptic and celestial equator intersect is due south of the star ω Psc.

In binoculars, κ Psc is an apparent double, consisting of unrelated stars of mag. 4.9 and 6.3. A somewhat similar double is formed by ρ Psc, a white star of mag. 5.4, and the orange giant 94 Psc, mag. 5.5.

TX Psc, also known as 19 Psc, is a very red star visible with the naked eye. It varies irregularly between mags. 4.8 and 5.2.

J	F	M	A	M	J
J	A	S	O	N	D

FACT FILE

SOUTH AT 22:00:
Nov.05

AREA:
(14th) 889 sq. deg.

DOUBLE STARS:
κ, ρ

VARIABLE STARS:
TX

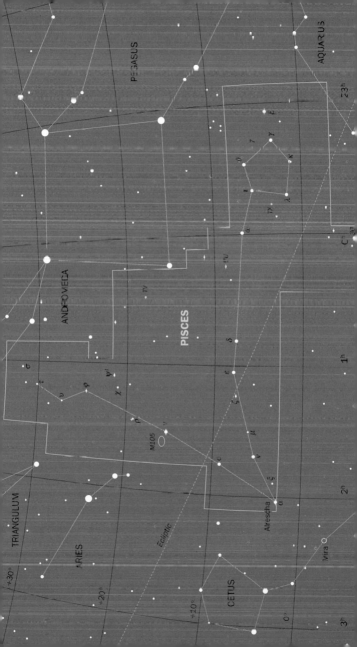

J	F	M	A	M	J
J	A	S	O	N	D

FACT FILE [Sge]

SOUTH AT 22:00:
Aug.20

AREA:
(86th) 80 sq. deg.

GLOBULAR CLUSTERS:
M71

FACT FILE [Vul]

SOUTH AT 22:00:
Aug.15

AREA:
(55th) 268 sq. deg.

NEBULAE:
M27

SAGITTA
Sagittae · Sge · The Arrow

VULPECULA
Vulpeculae · Vul · The Fox

These two small constellations lie in a very rich part of the Milky Way. On clear nights so many stars are visible in binoculars that it may sometimes be difficult to recognize the specific ones that you are trying to find.

Between γ and δ Sge lies M71, which is a globular cluster that is relatively poor in stars. North of γ Sge in Vulpecula is M27, the Dumbbell Nebula, visible as a hazy spot in binoculars. Binoculars also reveal α Vul, a red giant (mag. 4.4), as forming a wide double with 8 Vul, an orange giant (mag. 5.8).

The brightest gaseous nebulae

NGC 7000 (Cygnus)   M42 (Orion)   M7 (Sagittarius)

M11 (Sagittarius)   M27 (Vulpecula)

J	F	M	A	M	J
J	A	S	O	N	D

SAGITTARIUS

Sagittarii · Sgr · The Archer

FACT FILE

SOUTH AT 22:00:
Jly.30

AREA:
(15th) 867 sq. deg.

OPEN CLUSTERS:
M23, M24, M25

GLOBULAR CLUSTERS:
M22, M28, M55

NEBULAE:
M8, M17

This ancient constellation represents a centaur (half man, half horse), but unlike the southern constellation of Centaurus, Sagittarius has always been shown as an archer. The Sun is now in this constellation at the winter solstice, so it is most easily seen six months later, during the short nights of summer. Parts of the constellation, such as β Sgr, a naked-eye double, are visible only from southern Europe.

The centre of the Milky Way galaxy lies in Sagittarius, so it is exceptionally rich in stars and star clusters, with 15 Messier objects alone. The crowded star-fields and dark nebulae (intervening dust clouds) are a spectacular sight when examined with low magnifications.

M24 is small open cluster within an extremely dense star-field. Other open clusters are M25 (about 30 members), and M23, with more members, but at the limit for most binoculars. The most outstanding globular cluster is M22, visible to the naked eye and one of the finest in the sky. Two other globulars are M28 and M55, although the latter is rather faint even with binoculars.

Sagittarius also contains some prominent gaseous nebulae, notably M8, the Lagoon Nebula (which is just visible to the naked eye), and M17, the Omega Nebula. Both are clouds of gas excited to luminosity by embedded stars.

J	F	M	A	M	J
J	A	S	O	N	D

FACT FILE

SOUTH AT 22:00:
Jly.01

AREA:
(33rd) 497 sq. deg.

DOUBLE STARS:
ω

OPEN CLUSTERS:
M6, M7

GLOBULAR CLUSTERS:
M4, M80

SCORPIUS
Scorpii · Sco · The Scorpion

Scorpius is supposed to represent the scorpion that killed Orion, which is why it is on the opposite side of the sky, rising when Orion sets. It was originally a much larger constellation, including the area now known as Libra, which formed the claws. The Sun spends less time in Scorpius than in any other zodiacal constellation. Unfortunately the southern-most stars, representing the tail and sting are low, or below the horizon, for most of Europe.

As with Sagittarius, this is a crowded area of the Milky Way, and repays scanning with binoculars under clear skies. The most notable star is Antares ('rival of Mars'), a slightly variable, red supergiant about 400 times the size of the Sun that lies at a distance of 185 light-years.

M4 is a large globular cluster just west of Antares that does not have a strong central condensation and, although visible in binoculars, is hazy and not particularly easy to see. It is at a distance of 6,800 light-years. M80 is more condensed and probably as easy to spot, although it is much farther away at 27,000 light-years.

M6 (the Butterfly Cluster) is a rich open cluster in binoculars, which reveal some of the individual stars. M7, a little farther south, is so large that it is visible to the naked eye. Again, it is resolved with binoculars.

ω Sco is a naked-eye double: ω^1 is a blue-white star of mag. 3.9; ω^2, the closer star, is yellow and mag. 4.3.

FACT FILE

SOUTH AT 22:00:
Aug.01

AREA:
109 sq. deg. (84th)

VARIABLE STARS:
R

OPEN CLUSTERS:
M11

SCUTUM
Scuti · Scu · The Shield

This tiny constellation, in the rich star-fields of the Milky Way, was introduced in 1684 by Hevelius as Scutum Sobiescianum, Sobieski's Shield, in honour of his patron. One naked-eye star, δ Sct, is the prototype of a class of pulsating variable stars that alter with periods of a few hours. The variations are too small to be detectable visually.

There is one notable variable, R Sct, a red semiregular star, which changes between mags. 5.0 and 8.4, and shows characteristic double maxima (i.e., two maxima separated by a shallow dip, then followed by a deeper fade). It is regularly monitored by amateurs using binoculars.

M11, the Wild Duck cluster, has a noticeable fan shape (hence the name). It is one of the most concentrated of open clusters. An unrelated foreground star lies close to the apex of the fan.

FACT FILE

SOUTH AT 22:00:
Apr.25

AREA:
(47th) 314 sq. deg.

SEXTANS
Sextantis • Sex • The Sextant

Sextans (originally Sextans Uraniae) is another constellation that was introduced by Hevelius in 1687. Rather like Lynx and Leo Minor, it tends to be ignored by many astronomers. In fact it is so close to the ecliptic that the planets often cross its border for a short period. Its stars are even fainter than those of Lynx, however, with the brightest (α) no more than mag. 4.5.

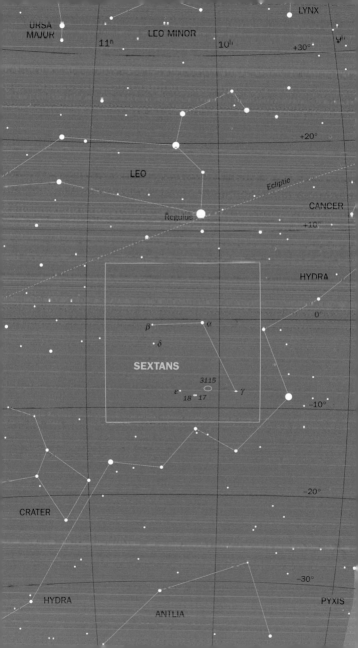

TAURUS

Tauri · Tau · The Bull

J F M A M J
J A S O N D

FACT FILE

SOUTH AT 22:00:
Dec.30

AREA:
(17th) 797 sq. deg.

DOUBLE STARS: ,
κ ϑ, σ

VARIABLE STARS:
λ

OPEN CLUSTERS:
Hyades, M45

METEORS:
Taurids

Taurus is a very ancient constellation and was linked with a bull in several different mythologies. It contains two remarkable open clusters, the Hyades and the Pleiades, as well as the giant star Aldebaran, which is so relatively close (65 light-years) and bright (mag. 0.9) that its orange coloration is easily visible.

The Hyades are so close to us that the stars appear well-separated even to the naked eye, giving rise to the 'V' of stars near Aldebaran (which is not a cluster member). Because they are so close, averaging about 150 light-years, their distances may be measured particularly accurately, and they are an important step in establishing the cosmic distance scale.

The Pleiades, M45, sometimes known as the Seven Sisters, are slightly farther away, between about 360 and 400 light-years. They

are a cluster of young blue stars about 78 million years old. Six of the seven main stars are normally easily visible to the naked eye, and some people are able to detect as many as nine. In fact the whole cluster contains about 500 stars. It is a magnificent sight in binoculars.

Taurus contains numerous double stars. ϑ is a naked-eye double: ϑ¹, white, mag. 3.8; ϑ², yellow, mag. 3.4, the brightest of the Hyades. κ¹ and κ² Tau are white stars forming a wide double, of mag. 4.2 and 5.3, respectively. The latter is sometimes known as 67 Tau, and both are members of the Hyades cluster as are σ¹ and σ², another wide double that consists of two white stars, respectively mags. 5.1 and 4.7. ζ Tau is an eclipsing binary star similar to Algol (p.230), and varies between 3.4 and 3.9, with a period of 3.95 days.

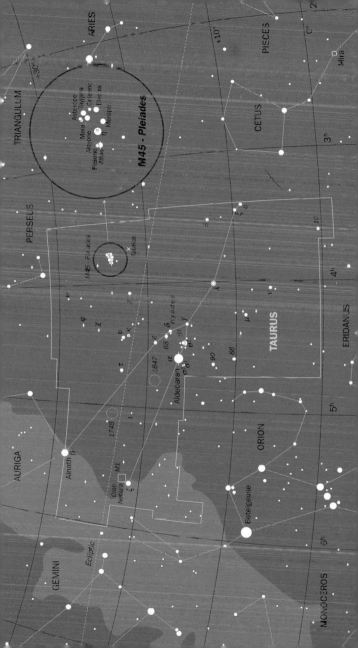

URSA MAJOR
Ursae Majoris · UMa ·
The Great Bear

FACT FILE

SOUTH AT 22:00:
Apr.15

AREA:
(3rd) 1280 sq. deg.

DOUBLE STARS:
ζ

GALAXIES:
M81, M82, M101

Many cultures have associated this constellation with a bear, and indeed the word 'arctic' comes originally from the Greek word for 'bear'. Most people, of course, tend to think that the constellation just consists of the Plough, the brightest seven stars, but in fact it extends over a very large area of sky. Because it is well away from the plane of the Milky Way, it contains many galaxies, but most of these are too faint to be visible without fairly large telescopes.

The most famous object is undoubtedly Mizar, ζ UMa (mag. 2.3), which keen eyesight will reveal as a wide double with Alcor (mag. 4.0). The two stars are at slightly different distances, however (78 and 81 light-years, respectively), so they are not a binary pair.

Mizar, however, is a true binary and has companion of mag. 4.0. By strange coincidence, these two stars and Alcor all prove to consist of extremely close binaries, so that the group actually contains six stars.

M81 is a spiral galaxy, just too faint (mag. 6.9) to be seen by the naked eye, but visible in binoculars. The same field also contains M82, which is often described as an irregular galaxy, but is actually an unusual spiral. At the other end of the constellation is M101, a third spiral galaxy, which is nearly as large as the Moon, but may be difficult to see with binoculars because of its low surface brightness.

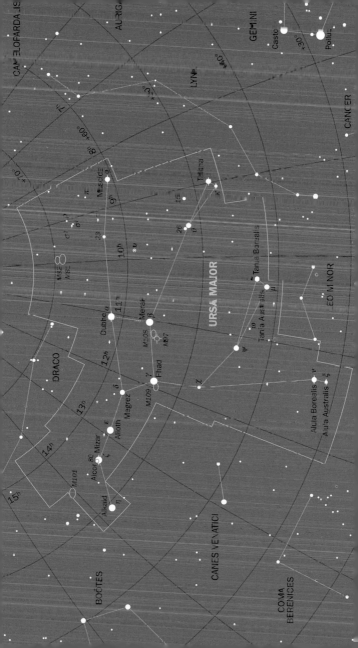

J	F	M	A	M	J
J	A	S	O	N	D

FACT FILE

ON MERIDIAN AT 22:00:
May.05

AREA:
(56th) 256 sq. deg.

DOUBLE STARS:
γ, η

VARIABLE STARS:
α

URSA MINOR
Ursae Minoris · UMi ·
The Little Bear

This constellation is believed to have been introduced around 600 BC by the gifted Greek astronomer Thales of Miletus. The position of Polaris so close to the North Celestial Pole made it an essential aid to navigation for early sailors and travellers. In the 16th Century the constellation became known as 'Cynosura' (derived from the Greek for 'dog with tail'), whence the saying that something is the cynosure of all eyes when it attracts constant attention.

Polaris, α UMi, is a yellow supergiant and a form of Cepheid variable (p.194). Its amplitude has decreased greatly in recent years, and it may be coming to the end of its variable stage.

There are two wide, apparent doubles, but in neither case are the stars true binaries. γ UMi, Pherkad, is a white giant (mag. 3.0), 480 light-years away. The naked eye or binoculars will reveal 11 UMi, which is an orange giant (mag. 5.0) that is closer to us (390 light-years).

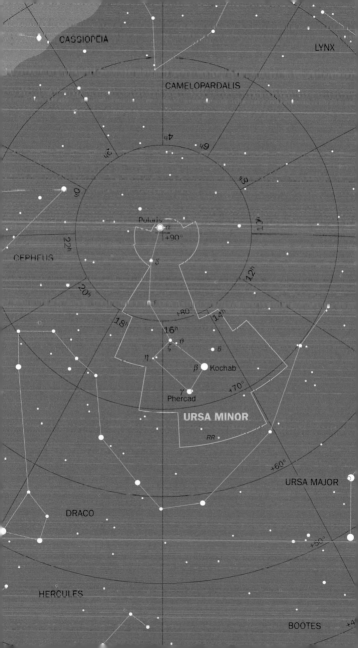

Virgo

Virginis • Vir • The Virgin

FACT FILE

South at 22:00:
May.05

Area:
(2nd) 1294 sq. deg.

Galaxies:
M49, M84, M86, M87

Virgo, the second largest constellation in the sky (after Hydra) has often been associated with a fertility goddess: to the Babylonians this was Ishtar; to the Greeks, Demeter; and to the Romans, Ceres (or Astraea, the goddess of justice). The name of the brightest star, Spica, α Vir, means 'ear of wheat'. Although it contains vast numbers of external galaxies in what is called the Virgo Cluster, the majority of these are too faint to show much detail with even the largest amateur telescopes. For such a large constellation, Virgo contains surprisingly few other objects of note.

Spica itself is a bright blue-white star at a distance of 262 light-years. It actually consists of two stars that are so close together that their shapes are tidally distorted, producing a slight (and visually undetectable) variation in brightness as they orbit one another.

A few galaxies are bright enough to be seen as hazy patches through binoculars. M49 is probably the brightest galaxy in Virgo. It is a giant elliptical system. M84 and M86 are close together and again both are elliptical galaxies. M87 is not quite as bright as M49, but it is a gigantic elliptical galaxy and actually one of the most massive known, containing perhaps 10 times as much material as our own (spiral) Galaxy.

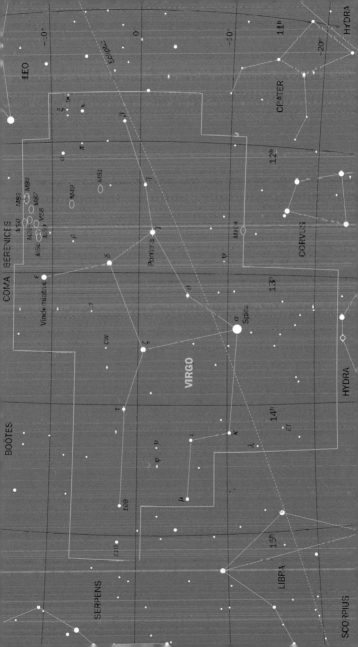

INDEX

WHAT NEXT?

The following books, magazines, CD-ROMs, and societies provide more information on astronomy and the night sky:

Books

Illingworth, V. (ed.), 1994, *Collins Dictionary of Astronomy*, HarperCollins*Publishers*

Karkoschka, E., 1990, *The Observer's Sky Atlas*, Springer-Verlag

Maris Multimedia, 1998, *Redshift 3*, CD-ROM, Dorling-Kindersley

Moore, P. (ed.), 1995, *The Observational Amateur Astronomer*, Springer-Verlag

Philips' Planisphere, Latitude 51.5°N, Geo. Philip & Son

Ridpath, I. & Tirion, W., 1993, *Collins Pocket Guide Stars & Planets*, 2nd edn., HarperCollins*Publishers*

Ridpath, I. (ed.), 1998, *Norton's Star Atlas*, 19th edn., Longman

Ridpath, I. (ed.), 1997, *Oxford Dictionary of Astronomy*, Oxford University Press

Rukl, A., 1990, *Atlas of the Moon*, Hamlyn

Tirion, W., 1996, *Cambridge Star Atlas*, Cambridge University Press

Tirion, W. & Sinnott, R., 1998, *Sky Atlas 2000.0*, 2nd edn., Cambridge University Press

Magazines (available in UK)

Astronomy, Astro Media Corp., Milwaukee

Astronomy Now, Intra Press, London

Sky & Telescope, Sky Publishing Corp., Cambridge, Mass.

Societies

British Astronomical Association, Burlington House, Piccadilly, London W1V 9AG

The leading organisation for amateurs, founded in 1890, with a particular emphasis on observational work. It publishes a bi-monthly *Journal*, which is recognized by amateur and professional astronomers worldwide for both popular articles and for the publication of accurate scientific research.

Society for Popular Astronomy, 36 Fairway, Keyworth, Nottingham NG12 5DU

An organisation intended for beginning amateur astronomers of all ages, particularly those without a great deal of equipment. It encourages amateurs to carry out simple observational programmes. Publishes a quarterly journal, *Popular Astronomy*, with readable, non-technical articles.

OTHER COLLINS WILD GUIDES

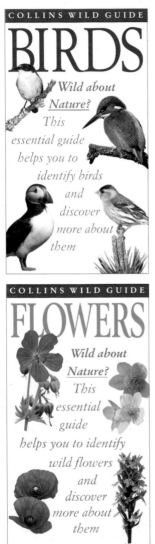

COLLINS WILD GUIDE

BIRDS

Wild about Nature? This essential guide helps you to identify birds and discover more about them

COLLINS WILD GUIDE

BUTTERFLIES & MOTHS

Wild about Nature? This essential guide helps you to identify butterflies & moths and discover more about them

COLLINS WILD GUIDE

FLOWERS

Wild about Nature? This essential guide helps you to identify wild flowers and discover more about them

COLLINS WILD GUIDE

MUSHROOMS & TOADSTOOLS

Wild about Nature? This essential guide helps you to identify mushrooms & toadstools and discover more about them